$2.50

For George Warner,
a fellow intelligence officer,
with best regards

Bob Evans

Christmas, 1980

LEGIONS OF IMPERIAL ROME

LEGIONS OF IMPERIAL ROME

An Informal Order of Battle Study

by

Robert F. Evans
Colonel, U.S. Army (Ret.)

Illustrations by
Katharine Evans

VANTAGE PRESS
New York / Washington / Atlanta
Los Angeles / Chicago

FIRST EDITION

Published by Vantage Press, Inc.
516 West 34th Street, New York, New York 10001

Manufactured in the United States of America
ISBN: 533-04618-1

Library of Congress Catalog Card No.: 80-50256

To Jane, my wife, a tactful editor

Contents

Acknowledgments

Many friends have helped in the preparation of this study, but special thanks are due to three in particular. John Headley, professor of history at the University of North Carolina at Chapel Hill, has made valuable suggestions and called attention to a number of errors in the first draft. Hunter Rawlings, head of the Department of Classics at the University of Colorado at Boulder, has given strong encouragement to the author and proposed a number of changes that have immeasurably improved the book. Maxwell Davenport Taylor, General, U.S.A. (Ret.), read and made helpful comments on the third draft.

Introduction

This study came about as the result of visits to places in Egypt, Tunisia, the Rhineland, and England where Roman legions had been stationed. The total ignorance or the massive misinformation concerning these legions on the part of the casually met native inhabitant inspired this work. It aims to present brief histories of individual legions during the first three centuries of our era, but it does not pretend to be a history of Rome, its emperors, or its wars.

The method used in preparing this study is that of the army intelligence service[1] rather than that of the scholar. It is, in effect, an order of battle study of Roman legions. Works of both ancient and modern authors have been studied for references to specific legions. These were then collated to produce individual legion histories. When conflicting reports turned up, and they often did, the most plausible was used without an extensive discussion of evidence. All works utilized are listed in the bibliography and are readily available.

No effort has been made to record the history of the auxiliary troops. The reader may assume that auxiliary cavalry and infantry units were attached to the legions in most combat situations.

Historical evidence concerning the legions is plentiful for the first century and a half, but for the period A.D. 150–300 it is sparse. The reader will note a concentration of incidents before the year A.D. 150. Histories of legions formed after the year A.D. 200 are not included, as these legions did not, in general, conform to the previous pattern and were often small organizations and "legions" in name only. Similarly, numbered but unnamed legions that did not survive A.D. 10 are not listed.

Geographical terms have been converted, where possible, to modern usage on the assumption that the reader is not fully informed concerning first-century names for countries, provinces, and cities. For example, St. Albans is used instead of Verulamium and northern Turkey instead of Pontus. With a modern atlas, the reader should be able to follow the tracks of the legions.

Legions are mentioned by number and name, but honorific additional titles, such as "pia fidelis," awarded to some seven legions at various times, are usually omitted except in describing the event bringing forth the honor.

The first chapter presents a general description of the legion as a fighting force. The next three chapters record in some detail the stories of nine legions. The trio in each of these chapters served together in at least one campaign and had common experiences. They were all in being when Augustus reorganized the Roman army at the turn of the millennium. One was probably destroyed 161 years later, five survived until the first half of the third century, one to the end of the fourth century, and two had shadowy existences in the beginning of the fifth. Major events described in these three chapters are, in subsequent chapters, mentioned but not redescribed.

The fifth chapter presents briefer histories of sixteen more Augustan legions. The sixth chapter contains accounts of nine[2] legions formed in the first century, and the final chapter covers the seven formed in the second century.

Five maps are included in this study to show the distribution of the legions in A.D. 30, 80, 130, 180, and 230. The base of the flag staff of a legion indicates its location. When no staff is shown, the legion has been placed in the general area, although its exact location has not been determined. A sixth map is included to show the Roman names for provinces and other areas during the second century.

The portrait drawings of the emperors by Katharine Evans are, in general, based on numismatic portraits executed during the lifetimes of the emperors.

Notes

1. Major modern army headquarters maintain order of battle sections that keep military histories of significant enemy units.

2. Omitted from this chapter is a legion raised by Clodius Macer, a governor of Africa who lost his life attempting to replace Nero as emperor. The legion sometimes called I Macrianna had a second flicker of life during the few months of the reign of Vitellius and then disappears.

LEGIONS OF
IMPERIAL ROME

CHAPTER I

The Nature of the Imperial Legions

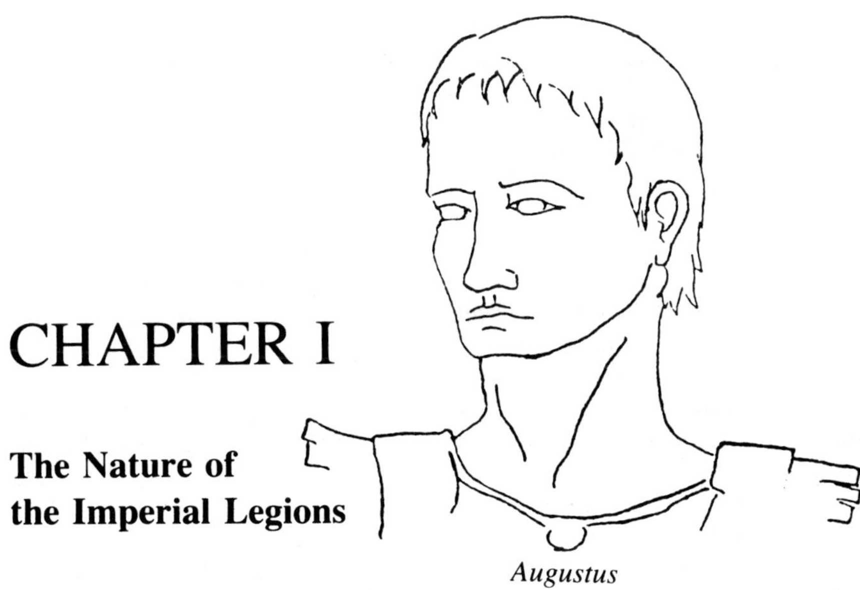

Augustus

Possibly the most effective military organization that the world has yet seen was the Imperial Roman Army, built around the legions.

The Emperor Augustus (27 B.C.–A.D. 14), profiting from the tactical and organizational genius of Caesar, Anthony, and their predecessors, established a Roman standing army of twenty-eight legions. This army, supported by auxiliary troops, was to remain unchanged in basic essentials for almost three centuries.

Until Augustus, legions had been recruited as needed and subsequently disbanded. The Augustan legion was, on the other hand, a permanent military unit made up entirely of Roman citizens who could retire normally after twenty years' service. On retirement, the soldiers received from the emperor grants of money or land or both and usually lifetime exemption from taxation for themselves and their wives.[1]

To support this permanent structure, Augustus, in A.D. 6, set up a military treasury that received the proceeds of two new taxes, a 5 percent inheritance tax and a 1 percent sales tax on auctions in Rome. Thus, the soldier no longer had to look to his commander for both pay and whatever retirement the commander might consider appropriate.

Legion commanders were selected by the emperor and were considered his personal representatives. They were titled "*Legatus*" and were usually young men of the senatorial class whose military experience was probably limited to serving on the staff of another legion commander. After a couple of years as a *legatus*, a man usually went on to another important post such as commanding an army or governing a province.

Most were, or became, senators, and at least eight became emperors. Legion command was considered more an executive and leadership position than one involving a profound knowledge of tactics and strategy. The commander had a staff of six tribunes, one of whom was also of the senatorial class, destined himself shortly to command a legion. The other five were probably of the equestrian class, had had experience with auxiliary troops, and would probably go on to be commanders of important auxiliary cohorts. All six tribunes handled administrative matters for the legion while the sixty centurions exercised command, directly subordinate to the *legatus*. Each of the ten cohorts that made up the legion was, in combat, under the command of a senior centurion.

The centurions were the life and survival of the legion. They were professional soldiers, probably rose from the ranks, and were charged with training, discipline, command in combat, and the general morale and well-being of the legion. Each normally commanded a century. Obviously, in early history, the century had had a hundred men, but this was soon reduced to eighty and so remained. The centurion carried a staff or swagger stick of grape vine as the symbol of his office and used it on occasion to beat recalcitrant soldiers. His pay during the Augustan period was 3,750 denarii per year, compared to 300 denarii for new recruits. His authority was almost limitless. He created officers junior to himself, personally conducted training, and awarded leaves of absence at his own discretion. This discretion included a payment to the centurion by the man going on leave. This unwise practice was eliminated by the Emperor Otho in A.D. 69. The pay of the centurion was then increased by an equivalent amount.

The senior centurion in the legion was known as the "*Primus Pilus*" and was really the tactical commander. He was very well paid, receiving, in the time of Augustus, four times the pay of other centurions. Emperors made sure that the *primus pilus* was protected against inflation, as one notes that by the time of the Emperor Domitian (A.D. 81–96), he was receiving 20,000 denarii as compared to the 15,000 denarii of sixty years earlier. Centurions moved and were moved from legion to legion in accordance with their own desires or the needs of the army and often went on to higher civilian and military jobs. As an example, P. Anicius Maximus, who, as *primus pilus* of Legion XII Fulminata in Syria about A.D. 70, became camp prefect (camp commander) for II Augusta in Britain and then camp prefect in Egypt for two legions in a joint camp at Nicopolis,[2] just east of Alexandria.

The training in a legion was continuous over the years, as most of the soldiers were professional and served for at least twenty years. To join a legion, a soldier had to be a citizen of Rome, though sometimes not too much proof was required. Training for combat was conducted with double-weight swords, shields, and spears and was very realistic, to include the construction and then the capturing of fortifications. Catapults were also much used in training, as in combat, and twenty-mile marches were regularly made with full armor and equipment. A phrase was used that combat was no different from training except for blood.

Good health was essential and was ensured by regular baths and good food. Each legionary base had its baths, and regular bathing in properly heated Roman baths continued through these centuries. With the fall of Rome, regular bathing went out in Western Europe, not to be resumed for fourteen hundred years.

Although in the second and third centuries meat and wine were part of the soldiers' diet, in the first century, it was made up primarily of soup, bread, vegetables, lard, and vinegar mixed with water. One is reminded that Christ on the cross says, "I thirst," and is given on a sponge vinegar, which he receives and then dies.[3] This was doubtless sour wine vinegar and water given from the rations of the Roman soldiers guarding the crucifixion area. That it was a gesture of compassion from a soldier to one he then admired is made more than plausible by the statement of the centurion in command at the moment of Christ's death "Truly this man was the Son of God."[4]

That the soldiers' diet was a good one[5] is exemplified by the fact that during those three centuries the average height of the soldier increased several inches. Also, the life expectancy of the retired soldier seems to have been excellent. Julius Valens, a veteran of II Augusta, died at Caerleon near Cardiff at the age of 100, his wife at 75. T. Flavius Virilis, a retired centurion from Britain, died at 70 in Algeria, having served in all three of the legions stationed in Britain as well as III Parthica and III Augusta.[6]

One of the most important qualities of the legion was its many capabilities. Every soldier was also an engineer, and the legion built roads, walls, bridges, theaters, and ships. They mined silver and gold, dug canals, and manufactured not only weapons but also pottery, shoes, and other needed items. These works were not on a small scale. Hadrian's Wall, which ran the seventy miles from Newcastle to Carlisle, was constructed largely of stone. It was built by XX Valeria Victrix, VI Victrix, and II Augusta over the three-year period of A.D. 122–124. Twenty years later, the same three legions built the thirty-nine-mile Antonine wall farther north. Much of Hadrian's Wall still stands today.

One strategically valuable legionary work was the canal dug in A.D. 58, connecting the Saone and the Moselle rivers in France so that troop movements to Germany could be made by water via the Zuider Zee.

In the first century, III Augustus built the road from Thevesta (Tebessa) in Algeria to Carthage in Tunisia, about 180 miles. This legion saw almost all of its service in North Africa where it is first identified in the year A.D. 6. It was disbanded 232 years later by Gordian III for failing to support him in his bid to become emperor. We are fortunate to have preserved some of Emperor Hadrian's words when he addressed the centurions of the legion and auxiliary troops during an inspection in A.D. 128 at Lambaesis. Needless to say, they are words of praise.[7]

One important aspect of the legions was their uniformity. A centurion from a legion in Spain transferred to one in Arabia would find on arrival an encampment identical to the one he had just left. It was laid out in a square, surrounded by a ditch and four walls, each with its gates and streets that ran to the center of the camp. There, in the center, was housed, or rather enshrined, the eagle of the legion. This standard was its totem, the symbol of its life, its history, and its greatness. Directly below the standard was the bank. All soldiers were required to bank part of their pay, to be returned to them when they retired or transferred. Similarly separate money chests were also maintained there for each cohort where was deposited one-half of the cash gifts made on various occasions by the emperors to the soldiers. The members of the legion made certain that their bank never fell into enemy hands. Next to the bank was the commander's house. He alone could have his wife and family with him. Here, in the central area, in their designated locations, were the storehouses, the infirmary, and the armory. Around them, in regular formation, were the troop barracks, each separate room housing eight men and ten such rooms making up a century. The centurion usually had an eleventh room to himself. There was no general mess, but each eight-man squad prepared its own meals at an established hour, using the rations issued to it supplemented by squad purchases. The centurions ran a tight schedule with no wasted time or motion.

Discipline was strict within the legion, and the centurions had almost complete control. Whole cohorts, however, might be punished, the most violent punishment, rarely used, being "decimation." In the year A.D. 21, a regular cohort in Algeria broke and ran from tribal raiders. The cohort was decimated; selected by lot, every tenth man was flogged to death. In the year A.D. 68 or 69 I Adiutrix, a relatively new legion made up of former sailors from the Roman Navy, was ordered decimated by the Emperor Galba when it refused to respect his authority as emperor.

During the three centuries when the legions were supreme, a number of changes took place that were necessary to conform with the changing conditions. None of them, however, altered the basic quality of the legion. Early in the period, the first cohort of each legion was increased to include five double centuries of 160 men each for a total strength of eight hundred men. The other nine cohorts remained at six centuries each or 480 men. To the ten cohorts was added a cavalry detachment of 120 horsemen, giving a total strength for the legion of 5,240.

Another important change was an increase of the years of service from twenty to twenty-four. By Hadrian's time, A.D. 117–138, this had crept up to twenty-five or twenty-six years, a change that did not go well with the troops. The number of legions also varied but gradually increased. In A.D. 37, the original twenty-eight had been reduced to twenty-five, but during Vespasian's rule there were thirty-one. In Trajan's time (A.D. 98–117), it was back to thirty. In A.D. 161 it was thirty-three, and in A.D. 165, again thirty. During Caracalla's time (A.D. 211–217), the number was again thirty-three. The variation in number resulted from the abolition of some legions for opposing an emperor and the creation by emperors of legions designed to be particularly faithful to them. One example is VII Galbiana, raised in Spain in A.D. 68, and brought to Rome by the new and short-lived Emperor Galba. It was then sent off to the Danube area and disbanded a few years later by the Emperor Vespasian. Despite these variations of the original twenty-eight Augustan legions, fifteen were apparently still going strong in the fourth century.

After the revolt of A.D. 89, Emperor Domitian made two important changes. He forbade the stationing of two legions at the same camp, and he limited the amount of money that could be held in the legion bank. Gallienus, A.D. 260–268, forbade senators of Rome from commanding legions or even serving with the army. These last steps were taken to prevent the legions from becoming a base for replacing the emperor of the moment. There was good reason for this thinking. Following the death of Nero in June A.D. 68, Galba became emperor. He was killed in January A.D. 69, and Otho wore the purple until the legions loyal to him were overcome, resulting in his suicide. Vitellius then took over in Rome, to be killed himself in December A.D. 69 after his forces lost the second battle of Cremona to legions loyal to the next emperor, Vespasian.

This second battle of Cremona, described in detail by Tacitus,[8] was a near disaster for the empire as it drew legions from all the frontiers, opening them to foreign invasion. In a night engagement, six legions supporting Vespasian, with the help of auxiliary cavalry, overcame six legions belonging to the Vitellian cause. These later were helped, unsuccessfully, by elements of three more legions brought over from Britain, II Augusta, XII Gemina, and XX Valeria Victrix. It is interesting to note that after the succession was settled, all the legions, not much the worse for wear, returned to their stations and continued their missions of defending the frontiers.

The British legions were not only unsuccessful at Cremona but also in two other efforts to establish their own chosen emperor. In A.D. 196 Albinus, a pretender to the purple, took II Augusta, VI Victrix, and XX Valeria Victrix to Gaul, but his forces were defeated at Lyons, where he lost his life. A century later, in A.D. 289, these legions gave Carausis, still another aspirant for the role of emperor, a victory in Gaul, but after he had acquired northeast France, he was assassinated by his first minister, Allectus. Constantinus, in A.D 296, brought Britain back under Diocletian's control.

The great period of the legions came to an end during the time of Diocletian (A.D. 284–305) who increased the number of legions to sixty while diminishing their effectiveness and eliminating the close support of cavalry to infantry. The volunteer quality of the army then diminished, and its increased size required the drafting of soldiers.

It should not be assumed that the legions during their heyday did all the fighting or that they were, without exception, successful in combat. There were throughout this period as many men serving in the auxiliary formations as there were in the legions, about 150,000 in each category, for a total armed force of three hundred thousand. The auxiliary troops were normally organized in five-hundred-man cohorts under Roman commanders. The troops usually were not Roman citizens, and they drew only half the pay of their opposite numbers in the legions. There was a large amount of auxiliary cavalry with a few formations as large as a thousand horsemen. It must be remembered, however, that the stirrup had not yet been invented, and, as a result, the horseman, without seated leverage, had not the combat capability that developed in the Middle Ages when the stirrup came from Asia to Western Europe. Some auxiliary units were made up of tribal soldiers under their own chiefs, but these were rare and a late development. Most were from Roman provinces such as Spain, Gaul, Egypt, or Britain. During most of the period, recruiting was no problem as, like the legion soldier, on retirement the auxiliary trooper received a bounty from the emperor and, what was more important, Roman citizenship. A bronze plaque was posted in Rome naming him as a citizen and a bronze copy sent to him in his place of retirement.

These auxiliary units manned the actual frontier outposts and put out the brush fires. The legions, backing them up, were only rarely called on and then only when there was a major invasion. Some auxiliary cohorts remained in place, successfully doing their jobs for literally centuries. Cohort IV Gallorum was at Chesterholm, a fort on Hadrian's Wall, in A.D. 213 and was still there in A.D 410. It is reported that Cohort IX Batavorum had been at Passau for almost four hundred years when, in the middle of the fifth century, the pay chest failed to arrive. Some soldiers who were sent to Rome to get the pay were killed on the way. Thus did Cohort IX Batavorum learn that Rome had fallen.

As to the legions always being successful, there were occasional exceptions. One unfortunate legion was XII Fulminata, which participated in the effort to take Jerusalem in November A.D. 66 and was chased almost all the way back to the coast by the Jewish forces. When Jerusalem was finally captured and destroyed in A.D. 70, five legions, with auxiliary units attached, were required to do the job. The shame, however, of XII Fulminata was not wiped out, and it was banished from Syria and stationed at Melitene on the Euphrates. One of the original Augustan legions, it was to distinguish itself later when, in A.D. 172, Marcus Aurelius led it over the Danube to break the power of the Quadi and again in A.D. 232 when the Emperor Severus Alexander used it in the invasion of Persia. At the very beginning of the period, three legions, XVII, XVIII, and XIX, were wiped out in Saxony, but five years later the Romans reconquered the area. These legion numbers were never used again. Although the legions did not win every battle, they won almost every campaign.

For three hundred years, some thirty legions, about 150,000 men, supported by 150,000 auxiliaries, maintained internal security and defended the 4,000 miles of frontier of an empire of 70 million people. This empire included England and Wales, all of Western Europe to the Rhine, Austria, Hungary, Rumania, Bulgaria, Yugoslavia, Albania, Greece, Turkey, Syria, parts of Armenia, Persia, and Saudi Arabia, Lebanon, Israel, Egypt, Libya, Tunisia, Algeria, and Morocco.

Notes

1. See Naphtali Lewis and Meyer Reinhold, *Roman Civilization Sourcebook II: The Empire* (New York: Harper, Torchbook edition, 1966), pp. 527–539, for retirement rights, including directives of Domitian and Constantine. Also, see David J. Breeze and Brian Dobson, *Hadrian's Wall* (New York: Penguin, 1976; rev. ed., 1978), p. 190.

2. See H.M.D. Parker, *The Roman Legions* (New York: Barnes and Noble, 1958), pp. 193–194.

3. John 19: 28–30, King James version.

4. Mark 15: 39, King James version.

5. See Breeze and Dobson, p. 186.

6. See Anthony Birley, *Life in Roman Britain* (London: Batsford Ltd., 1964, 1976), p. 46.

7. Addressing the chief centurions of the legion, Hadrian pointed out that despite the absence of a detached cohort and other troops, the centurions were as good as usual and deserved his commendation. He congratulated the legion cavalry not only for its spirit but also for the most difficult of all exercises, javelin throwing when clad in armor. The attached cavalry cohort won special praise for its rapid construction of a stone fortification and for its combatlike maneuvers. He praised the First Company of Pannonians for their throwing of the javelin and the lance and gave them largess. The hurling of stones with slings and close-knit maneuver won praise for the cavalry of the Sixth Cohort of Commagenians. See Lewis and Reinhold, pp. 507–509.

8. Tacitus, *The Histories* (New York: Penguin, 1975), III, 1–35.

CHAPTER II

The Legions in Roman Britain

Claudius

Six legions participated in the conquest, occupation, and defense of Britain. Of the four that were used by the Emperor Claudius in A.D. 43 for the invasion of the island, two, II Augusta and XX Valeria Victrix, were to remain stationed in Britain for almost four hundred years. The third of the invading legions, XIV Gemina, was moved to the Rhineland in A.D. 70, and the fourth, IX Hispana, had been withdrawn to Nijmegen, Holland, prior to A.D. 120. II Adiutrix came to Britain in A.D. 70 and was sent from there to the Danube eighteen years later. VI Victrix was brought to Britain by the Emperor Hadrian in A.D. 120 when he was establishing the northern frontier and beginning the construction of Hadrian's wall. It was to remain stationed in Britain for almost three centuries.

This chapter tells the stories of three of the legions mentioned above, but some discussion of the continuing impact of the legions after the decline of Rome seems appropriate. In A.D. 410, the Emperor Honorius wrote to the British that they must, in the future, look to their own resources for defense against German and Scandinavian raiders. By this time, the effective troops of the legions had undoubtedly been withdrawn, but the concepts of training and discipline remained. They carried over to the troops organized by the British rulers, and no doubt the retired legionaires helped in the maintenance of these concepts. Three British rulers were able, thus, to preserve at least a part of Roman Britain for another two hundred years.

The first of these rulers, Vortigern, claimed to be the grandson of the Emperor Magnus Maximus. He brought to Britain German troops to help defend the country. These revolted, and Vortigern spent years in fighting against them. The second was Ambrosius Aurelanis. According to the various interpretations of the very scarce records of these times, he came to power about A.D. 460, claimed imperial ancestry, and successfully fought off the Irish, the Picts, the Scots, and the Saxons.

There are very faint historical records of the last of the three, Artorius, whose family owned one of the great estates or villas. About A.D. 475, he took over command of the British forces and was so successful in holding back the various enemies that some historians credit him with fifty years of peace and prosperity, and at least one names him as the last Roman Emperor of the West. His greatest battle, about A.D. 495, at an unlocated spot called Mons Badon, brought about the defeat of a large Saxon army by a small British one. He is the historical figure on whom is based the legendary King Arthur.

Artorius was killed in battle about A.D. 515, and then, slowly, Roman Britain began to unravel. The cities, already in trouble because of the collapse of trade, fell one by one to the invaders. In A.D. 552, Salisbury fell; in A.D. 571 it was the turn of Gloucester, Cirencester, and Bath. By A.D. 600, London was gone, and British rule survived only in Wales and in one or two small kingdoms north of the wall that the Romans had sponsored as buffer states. One of these, Strathclyde, survived all invasions and in A.D. 1018 was absorbed into the Scottish crown. Its capital, the Rock of Dumbarton (Dumbarton means in Celtic the fort of the British), finds its name transported by Scottish immigrants to the capital of the United States and to the famous Harvard headquarters there, Dumbarton Oaks.

Caracalla

VI Victrix

VI Victrix may have been organized by Octavian in the second decade B.C. but is first firmly identified in Spain in A.D. 9. It is last reported in the "Notitia Dignitatis," the official Roman Army list, 401 years later in York, England.

In both A.D. 23 and 63, it is noted in Spain and was, in the latter year, the only legion in Spain. It was in A.D. 68 that Sulpicius Galba was proclaimed emperor in the legion's Spanish camp. Galba was an unusual emperor, for, as Plutarch remarks, the job sought him, not he the job. Galba was the wealthiest Roman citizen and was serving as governor of Spain at the time. He had had a distinguished military and governmental career but had reached the age of seventy-three.

Nero, in Rome, committed suicide on June 9, his position having become impossible, being deserted on all sides. Galba, still in Spain, was chosen by the Senate to be the next emperor. Before starting for Rome late in the summer, he recruited another Spanish legion, VII Galbiana, which accompanied him to Rome. He would have been wiser to have taken VI Victrix with him as well. When he got to Rome, he sent VII Galbiana on to Carnuntum on the Danube east of Vienna. He thus had no troops loyal to him personally, and when his old-fashioned virtues of thrift and discipline came to light in a Rome already corrupted by the years of Nero's reign, he became very unpopular. On January 15, A.D. 69, he was assassinated by troops from the pretorian barracks at the instigation of Otho, who then became emperor. According to Plutarch, only one man fought to protect Emperor Galba. Before the rush of the revolting soldiers, Galba's chair carriers and guards all fled.[1] A centurion, Sempronius Densus, first tried to turn back the rebels with his vine switch, the symbol of authority. This failing, he drew his sword and engaged them all until he was brought down by a cut under the knees. One would like to think that this man was a centurion of either VI Victrix or VII Galbiana.

The following year, Vespasian, who had become emperor after the very brief reigns of Otho and Vitellius, sent VI Victrix from Spain to Germany as part of the eight-legion force concentrated to put down the revolt of Civilis.

In the year A.D. 80, it is identified at Novaesium (Neuss), on the Rhine thirty miles north of Cologne. In A.D. 89, Antonius Saturninus, Roman governor of upper Germany, led a revolt against the Emperor Domitian. The legions of lower Germany remained loyal, and VI Victrix, which had been moved to Vetera near Xanten, six miles north of Cologne, participated in the defeat of the rebel forces. As a reward for this decisive action, it was honored with the new and additional title of *Pia Fidelis Domitiano,* as were three other legions, I Minervia, X Gemina, and XXII Primigenia.

About A.D. 120, the then Emperor Hadrian brought the legion from Vetera over to Britain. It was stationed at Newcastle, but its headquarters is shown very shortly thereafter at York. There had been uprisings in Britain, and Hadrian had determined to construct a wall from Newcastle to Carlisle to control this area. The wall was built by this legion and two others, XX Valeria Victrix, and II Augusta. VI Victrix built the east end of the wall and, in addition, constructed a temple at Newcastle to the Roman gods Neptune and Ocean.

When completed, Hadrian's wall served not as a fortification but more as a sentry walk. Manned by auxiliary troops, the wall was a base from which cohorts could move out and engage enemies to the north. Another wall, farther north, the Antonine, was built by the same three legions twenty years later but did not prove successful and was largely abandoned twenty years after its construction.

Widespread revolt appears to have taken place in A.D. 155–158 requiring heavy fighting by the legion. Reinforcements were brought from Germany in A.D. 158 as fillers for the three legions that had been seriously reduced in strength by the fighting.

In A.D. 196, Albinus, governor of Britain, took VI Victrix together with the other two legions to the continent in an effort to establish himself as emperor. After his defeat, the legions returned to Britain to find York and Hadrian's wall completely overrun. York was recaptured and had to be rebuilt by VI Victrix. The wall was restored by A.D. 205, and in A.D. 208 the Emperor Severus came to Britain to lead personally a punitive expedition against invaders from the north. Troops of the legion were used by the emperor in his northern expeditions, which continued until A.D. 211 when the emperor died at York and was succeeded by Caracalla, who had accompanied him to Britain. Caracalla conducted still another expedition, which was highly successful. From this time on till almost the end of the third century, Britain was relatively peaceful.

The next real break in the peace came in A.D. 287 when Carausius set himself up as emperor of Gaul and Britain. He was murdered by Allectus, who, in turn, was defeated in A.D. 293 when Constantius Chlorus invaded from the continent and brought Britain back into the empire. In this war, the northern frontier again was stripped of troops and again overrun from the north. Again, VI Victrix returned and had to rebuild its base at York and portions of the wall.

For the next seventy years, conditions were stable, and Britain was prosperous, but in A.D. 367 a combined assault by Scots, Picts, and Saxons once more overthrew the wall. Once more, York and the wall had to be restored after the Emperor Valentinian sent Theodosius with an adequate force to suppress the uprising. Again, Britain was prosperous and relatively peaceful despite disasters in other parts of the empire. Barbarians were, however, knocking at the gates of the continental provinces. The Emperor Honorius withdrew most of the combat troops from Britain in A.D. 403, including XX Valeria Victrix, leaving a shadowy VI Victrix still at York when Roman rule over Britain came to an end in A.D. 410. In that year, Honorius wrote to remaining authorities in Britain that they were entirely on their own and could look for no more help from Rome.

Magnus Maximus

XX VALERIA VICTRIX

Records for Valeria Victrix, one of the original Augustan legions, also cover over four hundred years. The first firm date, however, is A.D. 6 when it was part of Tiberius' army in Illyricum (Yugoslavia). It derives its name from a commander in the Illyricum wars, Valerius Messalinus.

By A.D. 13, it had been moved to the Rhine frontier and in A.D. 37 was stationed at Novaesium (Neuss), fifty miles north of Cologne. In A.D. 73 the Emperor Claudius used it, with two other legions, for the invasion and conquest of Britain.

In A.D. 61, it was the main element that finally crushed the British uprising led by Boudicca.[2] Earlier her forces had badly mauled IX Hispana at Camulodunum, causing some two thousand casualties, and had captured London. The victory of the Roman forces was made more difficult by the failure of II Augusta to follow orders and intervene at the critical moment. Its acting commander, the camp prefect, Poenius Postumas, killed himself when he found out how costly had been his failure to respond. XX Valeria was given the additional title of "Victrix" for its part in the campaign.

At least part of the legion participated, on the losing side, in the second battle of Cremona, forty miles southeast of Milan. Vitellius had been hailed emperor by the legions on the Rhine and had moved on to Rome. Vespasian had been proclaimed emperor in Alexandria and elsewhere, and his forces were moving on Rome when this decisive battle was fought at night with some fifteen legions involved. With Vespasian's victory, the troops of XX Valeria Victrix were returned to Britain with the legion stationed at Chester on the northern border of Wales. Under the command of Julius Agricola, a future governor of Britain, the legion helped suppress another British uprising in A.D. 69 and 70.

Eight years later, when Julius Agricola had become governor, he used his old legion in the conquest of northern Wales and the destruction of the rebellious Ordovices. In A.D. 84, it was probably the main element in the victory over the Caledonian tribes at Mons Graupius that put the Romans in control of lowland Scotland.[3]

In A.D. 122 or 123, the legion joined II Augusta and VI Victrix in starting the construction of Hadrian's wall between Carlisle and Newcastle. This tremendous military work was not completed till A.D. 128, but today it remains the largest structure surviving from the Roman world. Twenty years later, the same three legions built the shorter but less successful Antonine wall farther north.

As was mentioned earlier, Albinus, governor of Britain in A.D. 196, took all three legions from Britain to Gaul to support his effort to become emperor. Near Lyons, a decisive battle took place with the forces of Severus, who had legions from the Danube and Rhine frontiers. Albinus lost both the battle and his life, and the three legions from Britain were returned, apparently little depleted, to their former stations. The Emperor Severus at this time, with the hope of preventing similar events in the future, divided Britain into two provinces so that no governor could command three legions. Although historical records are sparse for the third and fourth centuries, the legion is noted at Chester in both A.D. 208 and 220.

XX Valeria Victrix appears to have made another incursion on the continent when the usurper Magnus Maximus, with an army from Britain in AD. 383, took over Gaul and Spain in addition to Britain. The Emperor Gracian was in Paris and fled to Lyon, where Maximus's cavalry caught up with him and executed him. Maximus, who had received a reluctant acknowledgment by Theodosius, overextended himself in trying to take over Italy and other parts of the empire. He got as far as Milan but had to turn west to meet Theodosius's army coming up from the Danube area. At Siscia, in Yugoslavia, southeast of Zagreb, his forces were defeated, and he tried to escape to Italy. He was caught and executed at Aquileia just west of Trieste. Once more, the legion finds itself returned to Chester.

About A.D. 403, XX Valeria Victrix makes its final departure from Britain. The Emperor Honorius was being besieged by Alaric at Astra near Turin, and his general, Stilicho, rounded up all available troops, including this legion. He arrived in time to save the emperor and soundly defeat Alaric on Easter Sunday at Pollentia, twenty-five miles southeast of Turin

Drusus

IX HISPANA

Unlike XX Valeria Victrix and VI Victrix, the record of IX Hispana is relatively brief and covers only 155 years. There are vague interpretations that may refer to it as early as 43 B.C. but the first clear reference comes in A.D. 6 when it was transferred from Spain, where it had obviously been long stationed, to Hungary. Three years later, it is shown at Siscia in northeast Yugoslavia southeast of Zagreb.

In A.D. 14, after the death of the Emperor Augustus and the accession of Tiberius, the legion, which was in summer camp with VIII Augusta and XV Apollinaris in western Yugoslavia, joined the other two in a mutiny. Nearby villages were looted as well as the town of Nauportus, present-day Vrhnika, 140 miles north of Fiume. The troops demanded shorter terms of service, claiming that some soldiers were being held in the ranks for thirty years. More pay was also demanded. IX Hispana appears in a moderating role when the men of the other two legions almost came to blows as to whether a certain centurion should be executed. Other centurions found hiding places, according to the detailed description by Tacitus,[4] except for one named Lucilius, who was killed by his men. He had been nicknamed "Another Please" for his habit of breaking his staff on a soldier's back and then shouting for another and still another.

To the rebel legions, Tiberius sent his son Drusus, who, despite many promises, was failing to make much headway until there took place what seens to have been, from Tacitus's account, an eclipse of the moon. The troops thought this an ill omen, and then a violent storm convinced them that they were in trouble. Drusus took advantage of this and had the ringleaders killed. Early winter storms then developed, and these helped bring the troops into line, and all three legions marched off to their more comfortable permanent winter quarters. Thus ended the mutiny.

Guerrilla warfare having broken out in the province of Africa (Tunisia), IX Hispana, apparently then in good shape, was dispatched in the year A.D. 20 to help restore order. It seems to have been fairly successful for in A.D. 24 it was returned to its old headquarters at Siscia, leaving III Augusta to hold the Tunisian frontier. Nineteen years later, it was used in the invasion of Britain where it was to stay for the remainder of the century.

In A.D. 43, with II Augusta, XIV Gemina, and XX Valeria Victrix, it was landed in Kent for the campaign in Britain. With the legions spreading out over Britain, IX Hispana was established at Lincoln, 130 miles north of London, by A.D. 48. Its occupation duties were interrupted in A.D. 60 by the very serious uprising led by Boudicca, queen of the Iceni. The queen's forces overran and wiped out the small garrison at Colchester fifty miles northeast of London and destroyed the colony of retired soldiers recently established there. IX Hispana hastened to intervene and was badly mauled by the rebels and took extensive casualties. The then governor of Britain, Suetonius Paulinus, had to abandon both London and St. Albans to the queen's forces. He regrouped, however, and the following year decisively defeated the rebels and destroyed their capability of further effort. London and Colchester, having been burned to the ground, were rebuilt, and IX Hispana moved from Lincoln to York, where it had settled by A.D. 62. It was brought up to strength by the bringing of two thousand replacements from Germany.

Gnaeus Julius Agricola, a one-time commander of the legion, served as governor of Britain from A.D. 78 to 84. He was the father-in-law of the historian Tacitus and got a very good press from him in his campaigns in Scotland. In A.D. 83, IX Hispana was attacked in its advanced Scottish encampment by a large force of rebels and was relieved only in the nick of time by Agricola's other forces. Needless to say, Tacitus gives a vivid account of the battle with the soldiers of IX Hispana making a sally from their base when they hear the approach of the relieving forces. The final battle, which brought peace for some years to the area, was fought the following year at Mons Graupius[5] where the combined highland forces were crushed.

After these campaigns, the legion went back to York, its permanent base, where its presence is noted in A.D. 86, 88, and 110.

Among the historians, there was some controversy about the future of the legion. Some believed that it was wiped out by A.D. 119 in a British uprising.[6] This theory is no longer generally accepted, and it is now believed that it was brought back from Britain to the continent by the Emperor Trajan in the early part of the second century when it was stationed at Nijmegen, Holland. From here it was transferred to the East and probably destroyed in Armenia in the war with Parthia in A.D. 161. After that date, there are no further mentions of IX Hispana. In its 150 years, it had served Rome in Spain, Hungary, Yugoslavia, Tunisia, Britain, Holland, and Armenia.

Notes

1. Plutarch describes the incident as follows in *The Lives of the Noble Grecians and Romans* (New York: Modern Library,) "Galba," p. 1284:

> Galba got into his chair and was carried out to sacrifice to Jupiter, and so to show himself publicly. But coming into the forum, there met him there, like a turn of wind, the opposite story, that Otho had made himself master of the camp. And as usual in a crowd of such a size, some called to him to return back, others to move forwards; some encouraged him to be bold and fear nothing, others bade him to be cautious and distrust. And thus whilst his chair was tossed to and fro, as it were on the waves, often tottering, there appeared first horse, and straightway heavy-armed foot, coming through Paulus's court, and all with one accord crying out, "Down with this private man." Upon this, the crowd of people set off running, not to fly and disperse, but to possess themselves of the colonnades and elevated places of the forum, as it might be to get places to see a spectacle. And as soon as Atillius Vergilio knocked down one of Galba's statues, this was taken as the declaration of war, and they sent a discharge of darts upon Galba's litter, and missing their aim, came up and attacked him nearer hand with their naked swords. No man resisted or offered to stand up in his defence, save one only, a centurion, Sempronius Densus, the single man among so many thousands that the sun beheld that day act worthily of the Roman empire, who, though he had never received any favour from Galba, yet out of bravery and allegiance endeavoured to defend the litter. First, lifting up his switch of vine, with which the centurions correct the soldiers when disorderly, he called aloud to the aggressors, charging them not to touch their emperor. And when they came upon him hand-to-hand, he drew his sword, and made a defence for a long time, until at last he was cut under the knees and brought to the ground.
>
> Galba's chair was upset at the spot called the Lacus Curtius, where they ran up and struck at him as he lay in his corselet. He, however, offered his throat, bidding them "Strike, if it be for the Romans' good." He received several wounds on his legs and arms, and at last was struck in the throat, as most say, by one Camurius, a soldier of the fifteenth legion.

2. See *Cassius Dio Roman History* (Cambridge, Mass.: Harvard University Press, 1924), LXIII, 1–12, for detailed report on Boudicca's defeat.
3. See Tacitus, *The Agricola and the Germania* (New York: Penguin, 1970), 28–38.
4. See Tacitus, *The Annals of Imperial Rome* (New York: Penguin, 1956), p. 46.
5. Mons Graupius is yet to be precisely located.
6. See Parker, p. 161. Also see Breeze and Dobson, p. 29.

Notes

CHAPTER III

Jerusalem and Alexandria

The campaign that resulted in the capture and destruction of Jerusalem in A.D. 70 is one of the most completely recorded of all Roman conquests, thanks to Josephus, the capable commander of the Jewish defense for a large part of the time. Josephus, after his capture, turned quisling and aided the Romans in every way he could. After the fall of Jerusalem, he went to Rome and there wrote his famous history of the Jewish war, which defends his every action. This volume also contains the most detailed report that has come down to us on the organization, training, and tactics of the Roman legions.

Five legions were used in the campaign, which started when Nero was emperor and ended when Vespasian wore the purple. Vespasian himself conducted the first part of this war, and when he went off to Rome, his son Titus, who was to succeed his father as emperor, completed it. Of the five legions, three are reported on in this chapter, V Macedonica, XV Apollinaris, and X Fretensis. The last mentioned was commanded during the campaign by M. Ulpius Trajanus, the father of still another future emperor. The other two legions, VI Ferrata and XII Fulminata, are covered in chapter IV.

Alexandria loomed large in the Roman world, and both V Macedonica and XV Apollinaris saw service there, not all of it pleasant.

Titus

V MACEDONICA

The footprints of V Macedonica can be followed for three hundred years in Eastern Europe and the Middle East. An Augustan legion, it was probably formed by 25 B.C. but is first specifically mentioned in Syria in 15 B.C., having come from Macedonia where it had been stationed the previous year and whence it drew its name. By A.D. 23, it had been moved north and is found in Bulgaria as part of the army of the Danube. From there, it was sent to Turkey.

About the year A.D. 63, Nero moved it from Pontus in North Turkey to Melitene in Eastern Turkey, a key crossing of the Euphrates. At Melitene, the Roman general, Corbulo, concentrated four legions, including VI Ferrata, III Gallica, and XV Apollinaris, for the successful Armenian campaign of that year. It was, in reality, more a show of force than a campaign. V Macedonica had still been at Pontus in A.D. 57 and 58 when X Fretensis participated in the capture of the Armenian capital at Artaxata.

In A.D. 67, the legion joined with X Fretensis and XV Apollinaris to make up with auxillary troops the army that Vespasian used in the opening of the Jewish war. He first captured Gamara, twenty miles south of Tyre, which he burned to the ground and where he killed all the adult population. Next came the turn of Jotapata where the historian Josephus commanded the Jewish forces. According to the account of Josephus, the siege was protracted, with the Romans suffering many casualties. Vespasian is reported to have used 160 projectile hurlers as well as a great battering ram to breach the wall. Josephus' troops made many sorties and gave X Fretensis a bad time invading its very camp in one of the sorties.

Japha, a town in the neighborhood, observing the success of the defenders of Jotapata, revolted against the Romans. Vespasian sent M. Ulpius Trajanus, the father of the future Emperor Trajan and then commander of X Fretensis, with one thousand horsemen and two thousand infantry to handle the situation. Titus, Vespasian's son, then followed with more horsemen and one cohort of infantry. The town fell after a six-hour assault largely because most of the defenders had been caught outside the walls by Trajan. These Jewish soldiers had been wiped out as they could not get back into Japha or escape the Romans. The males of Japha were killed and the females and children sold into slavery.

To the south, in Samaria, another situation arose with Samarian troops concentrated at Mt. Gerizim. To Ceralius, commander of V Macedonica, fell this problem, which he handled with three thousand infantry and six hundred cavalry. Surrounding the enemy force, he offered them terms, guaranteeing their safety, if they would throw down their arms. As they refused, he attacked and wiped out the entire force, which, according to a gross exaggeration of Josephus, numbered 11,600.

Back at Jotapata, the siege continued into July A.D. 67 until the forty-seventh day, when the Roman platforms overtopped the wall. A night assault was led by Titus and was entirely successful. The town was destroyed; two days later, Josephus was found hiding with some twenty others in a pit. He finally surrendered and was kindly treated by Titus and Vespasian. V Macedonica was then sent to winter at Caesarea, on the coast forty miles northwest of Jerusalem.

In the next campaign, the legion had a rather bad time at Gamala, just east of the sea of Galilee, where after the Romans had broken in through the walls, they were chased out again, and Vespasian was almost captured in the city. The siege was continued, and, finally, the enemy was forced into the citadel in October A.D. 67. There they were holding out until, again according to Josephus, there came up a great wind storm, which helped the Romans by greatly lengthening the trajectory of their spears and arrows while having the opposite effect on the defenders. There were no survivors, and Josephus says that more died by committing suicide than were actually killed by the Romans.

We next find the legion at Emmaus, ten miles west of Jerusalem, holding this approach to the capital while Vespasian with other troops secured the surrounding area. When news arrived of the death of Nero in June of A.D. 68, Vespasian delayed the attack on Jerusalem itself. He then turned over the campaign to his son Titus, who concentrated the legions at the capital, V Macedonica coming there from Emmaus. The siege was a long one, and platforms were built by the legions to overlook the city walls. V Macedonica's tower was undermined by the Jews and collapsed and had to be entirely rebuilt. Finally, having broken through the first wall, the legion found itself facing a second one, equally strong. At two in the morning, the standard-bearer of the legion, with a trumpeter and twenty men, by stealth got on top of this wall. The sound of the trumpet fooled the defenders, who ran, and then the other Roman troops followed into the city. There continued much hard fighting, but the city was finally captured and destroyed in September of A.D. 70. Vespasian had, in the previous year, become emperor.

After the destruction of Jerusalem, Titus moved with V Macedonica to Caesarea on the coast where the booty was secured. From there, he marched over the desert to Alexandria with the legion. He then took a boat to Rome while V Macedonica was dispatched to Bulgaria, where it was stationed at Oescus on the Danube. It was still there in A.D. 96 but by A.D. 108 had been moved to Troesmis, farther down the Danube in Romania near present-day Baila, only fifty miles from the Black Sea. It was there in A.D. 138, but in A.D. 167 it had been moved to Potaissa, 150 miles north of the Danube in central Transylvania, to act as garrison for what was then known as Dacia. Rome decided, in due course, not to hold this area north of the Danube, and V Macedonica was returned by A.D. 274 to Oescus.

In A.D. 296, Narses of Persia decided to take Syria from the Romans. Galerius, the Roman commander there, was defeated, and on his return to Antioch the Emperor Diocletian gave him something of a tongue lashing and told him to try again, but this time he was given V Macedonica as a key element in his force. Narses was badly defeated. When Ctesiphon fell, the wife and children of Narses of Persia were captured together with great booty. Mesopotamia was surrendered to Rome, and Persia recognized a Roman protectorate over Armenia.

This can be thought of as the last campaign of V Macedonica as a true legion. Shortly thereafter, Diocletian expanded the army to sixty legions and split V Macedonica into two legions, apparently of the same name but each less than half strength. The newer legion was sent off for service in Egypt.

XV APOLLINARIS

Etruscan Apollo

XV Apollinaris was named for Augustus' protecting deity and was probably formed by him when he was still known as Octavian, prior to the battle of Actium in 31 B.C. First stationing is indicated in southern Yugoslavia, and it seems to have been moved prior to A.D. 14 to Emona on the River Save, fifty miles northeast of Trieste. It participated in the revolt[1] of A.D. 14. In A.D. 37, it is reported as having moved again, this time to Carnuntum, now Hamburg, on the Danube, thirty miles east of Vienna. In A.D. 63, it was sent to Melitene on the Euphrates where, under the command of Corbulo, it joined with three other legions in an impressive Armenian campaign.[2]

After the Armenian campaign, it is next found in Egypt, helping to put down an insurrection at Alexandria. As a result of an incident in the amphitheater, as described by Josephus, the Jewish population of the city attacked the Greeks.[3] Finally, two legions, including Apollinaris, were ordered by the city governor, Tiberius Alexander, to overrun the Jewish area known as Delta. After a pitched battle, the legions killed thousands of Jews and looted and destroyed their homes. Oddly enough, again according to Josephus, the looting and killing stopped at once when Alexander gave the order.

Because of the revolt of the Jews in Galilee, Samaria, and Judea, Titus was dispatched by his father Vespasian to bring the legion from Alexandria to Ptolemais, on the Mediterranean coast twenty miles south of Tyre, where it joined V Macedonica and X Fretensis for the campaign to restore Roman rule.

First Gamara, about twelve miles to the east, was captured and destroyed together with the adjacent villages. The city of Jotapata was then put under siege, with Josephus commanding the defending forces. The siege took forty-eight days. In the final nighttime assault on the city, Titus led a small group from XV Apollinaris, which achieved complete surprise and ended the defense.

After this victory, the legion was sent to refit at Scythopolis on the west side of the Jordan thirty-five miles east of Caesarea where the other two legions were sent. In September, Vespasian concentrated his forces at Scythopolis and moved against Tarichaeae, on the west shore of the Sea of Galilee. Here he found that the defenders, when unable to save the town, took to their boats and continued to fight from them. Vespasian then had his troops build rafts, which he manned with his own forces. A sea battle resulted in complete victory for the Romans with great loss of life among the Jewish sailors and destruction of most of the Jewish craft.

In October, it was the turn of Gamala on the other side of the Sea of Galilee. As in other sieges, the Romans built towers, the most important one here being that of XV Apollinaris. Battering rams were used as well as heavy stone hurlers, and, finally, the wall was breached, and the Romans poured in. The defenders rallied on high ground and drove out the attackers. It was here that Vespasian himself was almost killed or captured.[4] The city was finally taken in October.

After Gamala, the legion went with Vespasian to join in the attack on Jerusalem. During this prolonged siege, XV Apollinaris, under the command of Titus Phrygius, built the tower facing the High Priests Monument. This tower was destroyed by the defenders and had to be built a second time. It was this legion's giant battering ram with an iron beak that broke the first section of the wall. The legion also helped in the monumental task of building a wall all around the city, in three days, according to Josephus. The city finally fell in September A.D. 70.

The legion was now moved to Caesarea again, where prisoners were sold into slavery and booty made secure. Then it was back over the desert to Egypt, with Titus.

From Alexandria, the legion was shipped to Carnuntum on the Danube, east of Vienna, whence it had come seven years earlier. It is reported still on the Danube in A.D. 93, and again in A.D. 110. In 115, the Emperor Trajan moved it permanently to the Middle East, stationing it at Satala in northeast Turkey, near the Euphrates, and about a hundred miles from the Black Sea.

When Alexander was emperor, it is reported once more still at Satala in A.D. 232 about 270 years after its formation. It had spent almost half its time in Europe and half in the Middle East.

X FRETENSIS

Tiberius

Fretensis has a dictionary definition of "straits," and the body of water referred to in this case is the straits of Messina where Octavian (later called Augustus) won a naval war against Sextus Pompeius. According to Parker, the legion had as an emblem on its standard the trireme.

Identifications for the legion cover a period of 250 years, the first being in the year A.D. 6 when it was serving as garrison troops in northern Syria. In A.D. 18, it was in winter quarters north of Antioch at Cyrrhus. In that year, in its camp took place the confrontation described in such detail by Tacitus[5] between Germanicus, commander in the east, and his unwilling subordinate, Piso. Shortly thereafter, Germanicus died presumably from poisoning. Piso, on returning to Tiberius's Rome, was accused and, according to several reports, killed himself when he saw how difficult it would be to clear his name.

Records show X Fretensis still at Cyrrhus in A.D. 23, but it was brought north for the Armenian campaign of A.D. 57–58 and participated in the capture of the Armenian capital of Artaxata. Tacitus says that the three legions used by the Roman commander Corbulo, which included III Gallica and VI Ferrata, could not provide, even with their auxiliaries, enough troops for a garrison, so the city was razed, though the inhabitants were spared. Tiridates, the Armenian commander, had fled with his troops to fight again another day.

Shortly thereafter, the legion is noted in Cilicia in southeast Turkey, but in A.D. 63 it is again back in what was then Syria. M. Ulpius Trajanus, future governor of Syria and father of the future Emperor Trajan, was in command of the legion when Vespasian organized his troops for the Jewish war at Ptolemais, present-day Akko, thirty miles south of Tyre on the Mediterranean. X Fretensis had been brought there from the Euphrates, and it was to winter twenty-five miles farther south at the port of Caesarea in Samaria. In A.D. 67, it was used in the capture of Tarichacae and Gamala on the west and east shores of the Sea of Galilee, after which actions Vespasian based it at Scythopolis, present-day Bet She'an, west of the Jordan River and twenty miles to the south. In the campaign to capture Jerusalem, which started in A.D. 68 and ended in September A.D. 70, the legion for a time was positioned on the Mount of Olives. After the fall of the city, the legion went on in clean-up operations in A.D. 72 to capture Machaerus east of the Dead Sea, in what is now Jordan. This is where, some thirty years earlier, the head of John the Baptist had been presented on a platter to Salome after her famous dance.[6] In A.D. 73, it was the turn of Masada on the west bank of the Dead Sea, in present-day Israel, where the last defenders all committed suicide.

Although Jerusalem had been destroyed, the legion was stationed near the site as a garrison. It was still there in A.D. 114 and in A.D. 120. In A.D. 130, the Emperor Hadrian ordered the building of a new city where Jerusalem had stood, and the legion certainly must have had a part in the construction. The new city was to be known as Aelia Capitolina, but the Jews revolted and took the new city by storm in A.D. 132. After a severe struggle, the city was captured again by the Romans in A.D. 135 and again destroyed. The legion seems to have survived these events, for it is still reported in Judea in A.D. 138, and in A.D. 232 a report shows it still at Jerusalem. The city had been again rebuilt and was still a Roman colony in A.D. 637 when the caliph Omar captured it.

Notes

1. See IX Hispana, above, p. 16.
2. See V Macedonica, above, p. 22.
3. See Josephus, *The Jewish War* (New York: Penguin, 1959), p. 163.
4. See V Macedonica, above, p. 23.
5. Tacitus, *The Annals,* p. 106.
6. Mark 6: 21–28.

CHAPTER IV

Armenia, Syria, and Persia

The three legions discussed in this chapter played key roles in the Armenian and Persian wars. They crossed and recrossed Syria, but the most dramatic of the campaigns was the drive down the Tigris and Euphrates rivers to the Persian Gulf. In A.D. 114, the legions available to the Emperor Trajan for this campaign were III Gallica, VI Ferrata, XII Fulminata, II Traiana, XVI Flavia Firma, IV Scythica, and X Fretensis. This chapter tells the stories of the first three of these legions.

III GALLICA

Elagabalus as Sun God

III Gallica is reported to have been organized by Julius Caesar and at that time given the number III. It is supposed to have fought as part of Antony's army in the defeat of the Parthians in 36 B.C. and is shown in northern Syria as III Gallica in 4 B.C. It was probably recruited in Gaul, as its name implies.

In A.D. 23, another report shows it in Syria, and in the time of Claudius (A.D. 41–54), it was one of the four legions making up the Syrian garrison. In the Armenian campaign of A.D. 58, the legion played a key part. Vologes I of Parthia had installed his brother Tiridates as king of Armenia, and the Roman general, Cnaeus Domitius Corbulo, was required to bring Armenia back under Roman control. Apparently, like other Syrian troops, III Gallica was in slack condition and had to be toughened up, which Corbulo did with harsh conditions and discipline. With III Gallica and two other legions, VI Ferrata and X Fretensis, he launched a successful campaign, capturing simultaneously three major fortresses, including Volandum north of Mt. Ararat. Tiridate massed his army before his capital of Artaxata, and, during an entire day, the two armies faced each other with III Gallica on the left of the Roman line. During the following night, the Parthian commander thought better of the fight and withdrew, leaving Artaxata to the Roman mercy, which was not much. The city was razed. Corbulo then concentrated his legions on the Euphrates.

Another Roman commander, Lucius Caesennius Paetus, with three legions then went to war with the Parthian army in Armenia and took a bad beating, being allowed to escape from a siege only under humiliating conditions. Paetus' troops double-timed to the Euphrates where they had the protection of Corbulo's legions. Corbulo was then given overall command of the Roman forces and conducted a not very important campaign, with the result that Tiridates was left as ruler of Armenia, but he had to go to Rome to be invested by the emperor.

In A.D. 68, III Gallica was ordered to the Danube front in Yugoslavia. It first defeated the invading Rhoxolani, and then, in the spring of A.D. 69, fought a very successful action against nine thousand Sarmatian cavalrymen who had overrun auxiliary troops and were pillaging the Hungarian countryside. The Sarmatians were not only overconfident but laden with loot. The weather was snowy and the ground soft and wet. The horses had bad footing, and the Sarmatians, once unhorsed, were unable, because of their heavy armor and lack of shields, to fight on equal terms. They were exterminated.[1]

By October, the legion had reached Italy, moving with other Danube army legions to support the cause of Vespasian against Vitellius. On October 24, the two armies fought a meeting engagement that started in the afternoon and carried through the night. Tacitus recounts that at dawn the men of III Gallica turned to hail the rising sun in the tradition of the East. Their opponents, not knowing the custom, assumed that they were shouting a welcome to reinforcements and fled.[2] The following day, Vitellius's supporters had withdrawn to a fortified encampment near Cremona. This was assaulted by Vespasian's supporters, and it was III Gallica that first broke through the defenses. Vitellius's troops then withdrew within the walls of Cremona. Thinking better of resistance, once the attack on the town had started, the defending legions surrendered and marched out to the taunts of the victors. Cremona was then sacked and much of the population seized to be sold into slavery. The city, set afire, burned for four days.

Vespasian's legions, under the command of Antonius Primus, pressed on to Rome where Vitellius, awaiting his doom, was murdered by revolting local troops on December 20. There was some looting and burning by the Vespasian forces in Rome, but III Gallica was not billeted there but at Capua to the southeast. The legion was rather rough on the local population, particularly the leading families.

From Capua, III Gallica was sent back to Syria, whence it had come three years before. The legion was again in action in A.D. 114, this time as part of the army that the Emperor Trajan used in his successful war against Parthia (Persia). It will be recalled that 150 years earlier III Gallica had engaged the Parthians. This time, Armenia and Mesopotamia were annexed, and the Parthian capital, Ctesiphon, on the Euphrates about fifteen miles south of Bagdad, was captured. Trajan had divided his army into two parts, one going by boat down the Tigris and the other by boat down the Euphrates and then marching overland for the capture of Ctesiphon. It is not known with which part III Gallica found itself. In the winter of A.D. 115–116. Trajan continued down to the mouth of the Tigris, establishing temporary Roman dominance at the top of the Persian Gulf.

In A.D. 120, the legion is back in Syria and again is reported there during the reign of Antonius Pius, A.D. 138–161.

The spotlight of history falls once more on the legion on May 16, A.D. 218, when it was stationed at Raphanaea, about ninety miles south of Antioch. On this date, with the rising of the sun, Elagabalus was proclaimed emperor in the legion's camp. The son of the deceased Syrian, Varius Marcellus, who had had a distinguished career and become a Roman senator, Elagabalus was at this time only fourteen years of age. He was a hereditary priest of the god Baal of Emesa. Said to be the last of the Antonines, he was completely under the thumb of his mother, Julia Soamias Bassiana, who had engineered the coup.[3]

With his troops under the command of his tutor, he advanced in forced marches on Antioch, where the Emperor Macrinus organized to meet him. An engagement was fought twenty-eight miles south of Antioch, resulting in the complete defeat of Macrinus, most of whose troops deserted him. He tried to escape and flee to Rome but was recognized, despite his disguise, and executed.

III Gallica did not remain faithful to Elegabalus but supported enemies of the new emperor. As a result, it was disbanded in A.D. 218. By A.D. 232, however, it had been reactivated at Raphanaea and was moved to Danaba to cover the Palmyra-Damascus road.

VI FERRATA

Antony

The "iron legion," VI Ferrata, was, in all probability, one of the legions that Antony used at the battle of Philippi, in October 42 B.C. It would not, however, have been known as "Ferrata" at that time.

It is also just possible that it was the legion referred to in Chinese history[4] whose men confronted the forces of the Han Empire in Sogdiana in 36 B.C. Sogdiana, now known as the Turkmen S.S.R., lies to the east of the Caspian Sea on the Oxus River in southern Russia. In 36 B.C., Antony was campaigning to the west and south of the Caspian, and it is possible that a legionary detachment was dispatched in a probing action to the east and north. If the Chinese report is true, it would be the only recorded instance when the forces of the emperors of Rome and China met. One speculates that after seeing each other's troops and sizing up the situation, the two commanders each decided that they were on too long lines of supply to take on a difficult-appearing new enemy.

The legion in 4 B.C., when Augustus was reorganizing the army, is reported as garrison troops in northern Syria. In A.D. 19, it is reported stationed near Laodicia, south of Antioch, and again there in A.D. 23. Under the Emperor Claudius, A.D. 41–54, it continued as garrison troops for Syria.

In A.D. 58, it was part of Corbulo's army that defeated the forces of Tiridates in Armenia and destroyed the capital, Artaxata.[5]

In November A.D. 66, according to Josephus, the commander of the legion, Priscus by name, was killed when XII Fulminata was chased back from Jerusalem almost to the coast by the defenders of that city. Although it is not recorded whether any small element of VI Ferrata was present, it may be assumed, for it is unlikely that its commander was merely accompanying XII Fulminata. In any event, VI Ferrata participated in the siege and capture of Jerusalem in A.D. 70, an action discussed in more detail in the preceding chapter. After the fall of the Jewish capital, the legion was used in A.D. 72 in the clean up operations, in particular the capture of Machaerus and Masada on either side of the Dead Sea. It was then moved north to Samsat (Samosata), on the Euphrates, 200 miles northeast of Antioch, where it may have been joined by the newly formed XVI Flavia Firma.

In A.D. 105, Trajan had determined to take control of all of what is now Jordan and of much of Saudi Arabia. The task was assigned to VI Ferrata with some auxiliary troops attached and was efficiently accomplished. In A.D. 108, the legion is reported as garrison troops at Bostra in northern Jordan. A great Roman road was built between the years A.D. 111 and 114 that ran from the Gulf of Acaba to Petra, Amman (Philadelphia), Bostra, and on to Damascus. It may be assumed that the legion played a major part in this construction.

The Roman influence was now so great in this part of the world that a Roman fleet was maintained in the Red Sea and emissaries came to Rome from India.

Because of the legion's experience in the East, it is probable that Trajan used it as a key element of his force that in A.D. 115 swept through what is now Iraq down the Tigris and the Euphrates rivers to Bagdad and to Ctesiphon and on to the Persian Gulf.

In A.D. 120, the legion is reported in or near Antioch, about A.D. 130, at Samosata with XVI Flavia, in A.D. 138 back in Judea, and in A.D. 200 in Palestine. The last report available is for the year A.D. 232 in the reign of Alexander when it was moved from Caparcotna, in Palestine, to Phoenice (Lebanon). It disappears some time between A.D. 235 and 284.

One particular centurion who served with the legion in the second century deserves mention, for he illustrates the interchange of officers between the legions. Petronius Fortunatus, said to have come from Africa, probably Tunisia, served several years with I Italica in Romania where he became a centurion. He was then transferred to VI Ferrata in Palestine. Then he went on to serve in eleven other legions as follows: I Minervia in lower Germany, X Gemina in Yugoslavia, II Augusta in Britain, III Augusta in Algeria, III Gallica in Syria, XXX Ulpia in lower Germany, VI Victrix in Britain, III Cyrenaica in Jordan, XV Apollinaris in eastern Turkey, II Parthica at Albano in Italy, and, lastly, I Adiutrix in Yugoslavia. After some forty-six years of service, he retired and presumably went back to his native Tunisia.[6]

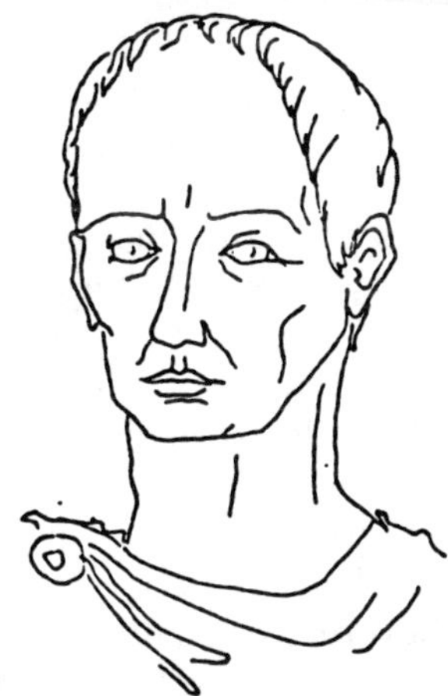

Julius Caesar

XII FULMINATA

A third legion that was part of Trajan's Middle East army when he undertook the Parthian campaign and moved down to the mouth of the Tigris was XII Fulminata. Another of the original Augustan legions, it may have had its origin with Caesar. Some state that it was once part of Antony's army. Its name means "hurler of lightning."

In 30 B.C., it is mentioned in Africa, probably Tunisia. By A.D. 14, it is reported in Syria and again in Syria in A.D. 23.

In A.D. 62, it was part of the unfortunate force led by Paetus into Armenia where he suffered a crushing defeat. The following year, it was sent back to Syria to refit. Again, in A.D. 66, it was humiliated, for it was used by C. Cestus Gallus in an abortive effort to capture Jerusalem. Although supported by auxiliaries and some elements from other legions, it was chased all the way back to Antipatris, forty miles away. It lost in the retreat its eagle and all its siege train and other equipment. It partially redeemed itself, however, in the forthcoming siege and final capture and destruction of Jerusalem in A.D. 70. Titus did not, however, return it to Syria but sent it to Melitene on the Euphrates River two hundred miles northeast of Antioch. It was in Cappadocia in eastern Turkey in A.D. 114 when Trajan undertook his campaign against Armenia and Mesopotamia and his expedition down to the Persian Gulf. It is not known whether XII Fulminata was part of the western force that went down the Euphrates, possibly as far as Babylon, before crossing overland to the Tigris or whether it was part of the eastern force that descended the Tigris past Baghdad all the way to the Gulf.

In A.D. 138, it is identified again in southeastern Turkey, but in A..D. 172 Marcus Aurelius led a force, including the legion, across the Danube into the lands of the Quadi, a people living in what is now northern Austria and Czechoslovakia. His base was the legion camp at Carnuntum east of Vienna, near Hamburg, and from this base, for three years, he relentlessly pursued the Quadi and their allies. It was during this campaign that he started writing his famous *Meditations*.

There is a story from this war, often repeated, of a victory of XII Fulminata in Moravia when a miraculous rainstorm put the enemy to flight. This was attributed to the prayers of the Christian soldiers of this Cappadocian legion.

By A.D. 231, the legion had been moved to Melitene on the Euphrates in eastern Turkey. The last campaign in which reports show XII Fulminata playing a part is the Persian war of A.D. 232. The then emperor, Alexander Severus, concentrated his legions in the general area of Antioch where he came himself, with his mother, to lead the forces against the Persian king of kings, Ardashir, who was bent on reestablishing the old powerful Persian Empire.

The campaign was no success for either the Romans or the Persians. The emperor advanced eastward in three columns. Ardashir first brought the northern column to a halt and then turned south and defeated the southern column. The emperor was presumably with the central column that never did engage the enemy due to the cautiousness of Severus.

Ardashir also must have suffered serious losses, for he broke off contact and withdrew eastward while the Romans withdrew to the Antioch area. It may be assumed that XII Fulminata was then returned to its base at Melitene.

Notes

1. See Tacitus, *The Histories,* 1, 79.

2. See Tacitus, *The Histories,* 3, 24–25; also Dio, LXIV, 14.

3. See *Cambridge Ancient History* (Cambridge: Cambridge University Press, 1939), vol. XII, p. 84.

4. See L. Carrington, Goodrich, *A Short History of the Chinese People* (New York: Harper Torchbooks, 1963), p. 52.

5. See III Gallica, above, p. 32.

6. See Graham Webster, *The Roman Imperial Army of the First and Second Centuries A.D.* (New York: Funk and Wagnalls, 1969), p. 119.

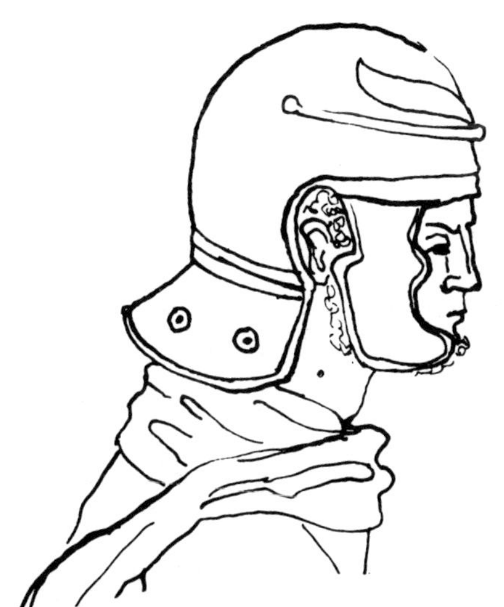

CHAPTER V

More Augustan Legions

When the Emperor Augustus died in the year A.D. 14, he bequeathed to the Roman Empire an army containing twenty-five legions. Of these, nine have been discussed in the three previous chapters, and the movements of the other sixteen are outlined below in numerical order.

The locations of these sixteen between the years A.D. 30 and 130 illustrate the changing pressures on the empire. As the maps on pages 101 and 103 show, an important shift of the legions to the east and south was taking place. In the year A.D. 30, there were eight legions on the Rhine, six in the general area of the Danube, and four in Turkey, Syria, and Palestine. One hundred years later, there were only five on the Rhine, but the six on the Danube had increased to ten and the four in the Middle East to eight.

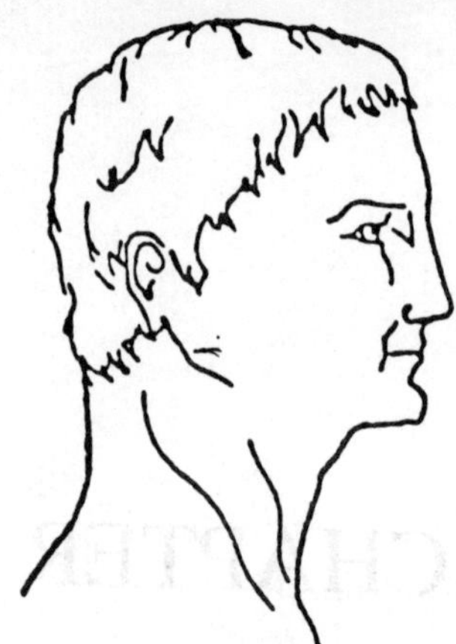
Germanicus

I GERMANICA

First noted as part of the army of Octavian before he assumed the title of Augustus, this legion was early in Spain where it suffered a disastrous defeat. It was subsequently reconstituted by Tiberius and stationed in the year A.D. 14 in Germany in a joint camp with XX Valeria Victrix at Cologne. That year, it participated in a revolt, and Tacitus reports, in detail, the successful personal appeal of Germanicus to persuade the troops to return to their loyalty.[1] Germanicus, who was then in command of the Roman armies in both upper and lower Germany, spent much time with the legion, and it may well be that from these events it derives its name. Apparently given to revolts, the legion joined the uprising of Civilis in A.D. 69. Civilis, a Batavian chief, persuaded the legion to take an oath to support his hoped-for Gallic empire. He posted the legion at Trier, the center for the Treveri tribe. When the army of the Emperor Vespasian arrived from Rome the following year, the Treveri went over to him. The legion abandoned Civilis and took the oath to Vespasian. Unmollified, Vespasian, in either A.D. 70 or 71, disbanded I Germanica.

II AUGUSTA

Capricorn as Standard

This legion is best known for its work in the construction of Hadrian's Wall. It was organized by Augustus, had the capricorn as emblem, and is first reported in Spain, whence, in A.D. 10, it was moved to the Rhine frontier. It is found on the Rhine at Strasbourg in A.D. 23. Twenty years later, the Emperor Claudius used it, then under the command of the future Emperor Vespasian, and three other legions for the invasion of Britain. Records show it was stationed at Gloucester in A.D. 62, the year after its then acting commander committed suicide when he learned the effect of his failing to obey orders during the uprising led by Boudicca.[2]

In A.D. 69, elements were sent to the continent to help defend Vitellius from Vespasian's drive for Rome. These elements were on the losing side at the second battle of Cremona and were promptly returned by Vespasian to the legion in Britain. In A.D. 75, the legion was moved to Caerleon in Wales where it was to be based for the indefinite future. In A.D. 122 and 123 it constructed major parts of Hadrian's Wall between England and Scotland and twenty years later helped with the construction of the short-lived Antonine wall farther north. In A.D. 196, the legion was brought to France by the pretender Albinus, but when Albinus was defeated by Severus, it was returned to Caerleon. In A.D. 383, it made one more trip to France with XX Valeria Victrix when Magnus Maximus made his unsuccessful effort to take over the empire. After the defeat of Maximus at Milan, the victorious Emperor Theodosius once again returned it to Caerleon. It was probably still in Britain in the beginning of the fifth century, though much diminished in effectiveness.

Hadrian

III AUGUSTA

This legion was not a world traveler. Its origin is uncertain, but it was probably formed and given its number by Octavian and given its name by him when he became Augustus. It is first reported in the year A.D. 6 at Ammaedara northeast of Tebessa in Algeria. From there it built the famous military road to Carthage. Reports for A.D. 23 and 70 show it still in Africa, presumably at Ammaedara. About the year A.D. 100, it helped construct the city of Thamugadi 100 miles to the west. This was undertaken at the direction of the Emperor Trajan. In the year A.D. 120 the legion is reported some twelve miles farther west at Lambaesis, near Batna, south of Algiers. It was here that the Emperor Hadrian inspected and addressed the troops in A.D. 128.[3]

The legion supported Severus in his drive to become emperor, and, as a result, it received the additional title in A.D. 194 of *piae vindicis*. This emperor is reported to have visited its camp at Lambaesis in A.D. 203. In A.D. 216, the legion was used by the Emperor Caracalla in his Parthian campaign in Mesopotamia.

In A.D. 238, back at Lambaesis, it supported the then emperor, Maximinus, against the rise of the Gordians. In that year, its troops killed in battle Gordian II, and Gordian I, as a result, killed himself. Gordian III, when firmly established as emperor, understandably disbanded the legion. It was, however, reconstituted, still at Lambaesis, in A.D. 253.

Aurelian

III CYRENAICA

The origin of the legion is very obscure. A part of Augustus's original army organization, it must have either come from, or at an early time served in, Cyrenaica in eastern Libya. The first firm report is for the year A.D. 7 when it was in Egypt, either at Thebes or at Coptos north of Luxor on the Nile. Other reports of it in Egypt are for the years A.D. 11, 23, 70, and 110. In A.D. 119, it was in the same camp as XXII Deioteriana at Nicopolis, Egypt. It was sent, probably in A.D. 120, to Arabia and stationed at Bostra in northern Jordan near present-day Irbid.

In A.D. 216, it participated in Caracalla's Parthian campaign in Mesopotamia but was back again in Bostra in A.D. 232.

It is likely that the legion was part of the force used by the Emperor Aurelian in the defeat of Zenobia and the capture of Palmyra in A.D. 272. The beautiful and forceful Zenobia, queen of Palmyra, had gained control of much of Asia Minor, including Syria and Egypt. She styled herself as queen of the East. Aurelian, determined to restore the unity of the empire, engaged her army first at Antioch and again at Emesa, winning both engagements. Zenobia fell back on Palmyra, but later she attempted to escape the seige of her capital. She was captured on the bank of the Euphrates and taken to Rome to be displayed in the triumph of Aurelian. Afterward she was well treated and lived in Rome with her sons in the manner of a prosperous Roman matron.

The last report we have of II Cyrenaica concerns its looting the temple of the sun god in captured Palmyra.

Galba

IV MACEDONICA

Caesar may have formed this unit, but certainly it was part of Octavian's army when he became Augustus. It was probably at an early date stationed in Macedonia. It had a short and not particularly distinguished career. First noted in Spain in the year A.D. 9 and again there in A.D. 23, it was brought to the Rhineland by Claudius after A.D. 39 to replace a legion that had been used in the invasion of Britain. It is reported sharing a double camp at Mainz with XX Primigenia in A.D. 43 and again in A.D. 69. It participated in the revolt against Galba, failed to support Vespasian, and, as a result, was disbanded by the latter in A.D. 70. At this time, Vespasian created two new legions, one of them being IV Flavia Firma, which might have received some of the better veterans from the defunct IV Macedonica.

Caligula

VII CLAUDIA

This legion at first had an unofficial name "Macedonica," presumably because at an early time it garrisoned Macedonia. It received its formal title and the honorific *pia fidelis* (patriotic and faithful) from the Emperor Claudius because of its loyalty when the troops in Dalmatia were urged to revolt in A.D. 42 by the governor, Furius Camillus Scribonianus. The legion had first been reported in this part of Yugoslovia in A.D. 39 during the reign of Caligula. In A.D. 57 or 58, it was moved up to the Danube. Like other legions on the Danube, it was drawn into the struggle in A.D. 69 between the forces of Vitellius and those of Vespasian. Fortunately, it was on the winning, Vespasian, side of the battle of Cremona, reportedly holding the left of the line in the first part of the battle. Afterward, returning to the Danube, it found itself posted to a new location at Viminacium, near Kostolacz, about forty miles down the Danube from Belgrade, Yugoslavia. Reports show it continuing there in A.D. 7, 86, 93, and in 108. In A.D. 115 a portion of the legion was used by Trajan in his Armenian campaign, but by A.D. 120 it is returned again to Viminacium. In A.D. 202, it is still there, and the last report is for A.D. 337 when it had an element in upper Moesia, probably at Viminacium.

Gallienus

VIII AUGUSTA

Bearing the name of the emperor this legion was established by Augustus and is first reported at Poetovio, near Marburg, on the Drava River in Yugoslavia and in the same area in A.D. 23 and 30. It took part in the short-lived rebellion in A.D. 14. About A.D. 43, it was moved to the Danube area of Moesia, probably Bulgaria. In A.D. 69, it went with the other two Moesian legions to Italy to support the cause of Vespasian at the battle of Cremona. Next, in A.D. 70, it was sent to the upper Rhine and stationed at Strasbourg. Reports show it continuing in the upper Rhine area in A.D. 83, 90, 110, 120, and 138. In A.D. 200, it is reported in lower Germany, and the last report is of its being used in A.D. 258–259 by the Emperor Gallienus in crushing a German invasion of Italy at a major battle at Milan.

Probus

X GEMINA

As its name implies, this legion was made up of two other legions. It is first reported in Spain in the year A.D. 9. Another report shows it in Spain in A.D. 23, but in A.D. 63 it was in Pannonia, probably Hungary. It returned to Spain in A.D. 69, but in A.D. 70 it was brought to Germany to help suppress the Civilis uprising. It arrived just too late for the battle of Vetera, and it was then established on the lower Rhine near Nijmegen.

During the time of the Emperor Domitian, A.D. 81–96, it helped in the suppression of another revolt and was awarded by the emperor the title *pia fidelis Domitiana*. About A.D. 108, it moved from Nijmegen to Vienna where it is reported in A.D. 120 and again, during the time of the Emperor Pertinax, in A.D. 193. In A.D. 196, it won another honorary title, *pia fidelis Severiana,* for having supported the claim of the new Emperor Severus. A report for the year A.D. 200 has it still in Vienna, and a last report for the year A.D. 260 shows it in upper Pannonia, presumably Vienna.

According to the often unreliable *Augustan History,* the legion could claim as former commanders two emperors, Aurelian and Probus.

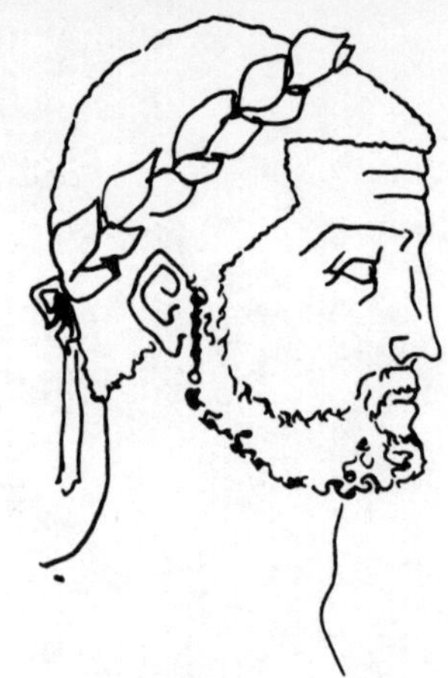

Diocletian

XI CLAUDIA pia fidelis

In pre-Augustan days, this legion was sometimes called Actiacus, which might imply fighting in the battle of Actium. It received its formal title in A.D. 42 for remaining loyal to the Emperor Claudius when a revolt had broken out under the leadership of Furius Camillus Scribonianus, governor of Dalmatia. Like VII Claudia, it received the honorific title of *pia fidelis*. In the year A.D. 9, it had first been noted in Dalmatia at Burnum near present-day Knin in Yugoslavia. It was reported there in A.D. 14 and 39 but in A.D. 70 moved to Switzerland and was stationed at Vindonissa, present-day Windisch, twenty miles east of Basel. After A.D. 101, but before A.D. 106, it was moved to Silistra (Durostorum) on the Danube in Bulgaria. Here it was reported in A.D. 110, 120, 138, and again in A.D. 200.

The last mention of the legion had it based at Aquileia in northeast Italy near Monfalcone. From here, Diocletian dispatched, in A.D. 288, a major element for the campaign in North Africa. This element participated in the resecuring of this part of the empire in a drive from Tangier to Carthage.

Vitellius

XIII GEMINA

A composite legion, put together by Augustus (Gemini means twin), XIII Gemina was reported on the Rhine at Windisch in Switzerland not far from Basel in the A.D. 9. It continued on the Rhine in this area in A.D. 14 and A.D. 23 but in A.D. 45 or 46 was sent to Poetovia, present-day Ptuj, near Marburg in Yugoslavia. By A.D. 96 it had moved to Vienna, but in the meantime, in A.D. 69, had been to Italy and fought on the Vitellian side in the first battle of Cremona and on the Vespasian side in the second. Trajan used it in his Dacian war of A.D. 106, after which, by A.D. 108, it was stationed at Apulum in Dacia near present-day Alba Iulia in Romania. It was again reported there in A.D. 120. In the next century, in A.D. 261, it was reported at Mehadia near present-day Balta in south-central Romania. Seven years later, it was back again at Poetovia in Yugoslavia. In A.D. 274, it was reported at Ratiaria near present-day Archer on the Danube in Bulgaria. After A.D. 300, it was reported split permanently into four elements, all stationed at different places on the Danube. It then ceased to function as a legion.

Otho

XIV GEMINA

This is another of Augustus's composite legions and first appears in the year A.D. 9, having been badly beaten in the Varian[4] disaster and reaching safety at Vetera, present-day Xanten in Germany on the lower Rhine. In A.D. 14, it was located in Germany, at Mainz as in A.D. 23. In A.D. 43, Claudius used it in the invasion and conquest of Britain, and it was one of the legions that finally crushed the uprising led by Boudicca in A.D. 60. For its part in this campaign, it was awarded the additional title of *Martia Victrix*. In A.D. 69, major elements were drawn to Italy to support Otho at the first battle of Cremona and were sent back to Britain by the winner, Vitellius. The following year, it was brought back to the continent by the next emperor, Vespasian, to help put down the uprising of Civilis at Xanten. Vespasian stationed it at Mainz. In A.D. 89, it was on the wrong side of a revolt against Domitian but was permitted to stay at Mainz for three more years. In A.D. 92 it redeemed itself, when XXI Rapax was destroyed on the Danube. Domitian sent for XIV Gemina, which put things to rights. It was then stationed in Pannonia (Hungary) and is noted there in A.D. 110 and in A.D. 120 at Carnuntum, east of Vienna on the Danube near Hainburg. In A.D. 236, it was still stationed in Pannonia at Carnuntum. The last report is from the year A.D. 260 when it remained in upper Pannonia, presumably at Carnuntum.

Tiberius

XVI GALLICA

This legion has a brief history: it seems to have had a genius for picking the wrong side. Its name appears to indicate service in Gaul before the first century, and the first location for it is upper Germany in A.D. 14. That year, the year of the death of Augustus, the legion participated in the short-lived mutiny against Tiberius. By the year A.D. 43, it was stationed at Mainz and before A.D. 69 had been transferred to Neuss (Novaesium), farther down the Rhine near Dusseldorf. The legion made the mistake of siding with Vitellius against the Vespasian forces at the second battle of Cremona. Even worse, the base elements of the legion that had been left behind at Neuss surrendered to the rebel leader Civilis and were marched in shame from Neuss to Trier. Shortly thereafter, probably in A.D. 70, Vespasian ordered the disbanding of the legion. He then organized a new legion in Asia Minor with the same number but with the name "Flavia firma," Flavia being his family name.

Domitian

XXI RAPAX

Another legion with a relatively short history, ninety-eight years, is XXI Rapax. It derived its name from *rapax,* meaning rapacious or greedy in the sense of overcoming all. First identified in the year A.D. 6 in southern Germany, it moves to Xanten on the lower Rhine. There it joined the other legions in a short-lived uprising in A.D. 14. In A.D. 45–46, it was moved to Windisch in Switzerland, near Basle, to replace XIII Gemina, which had been sent to Poetovio in Yugoslavia. For several years before this move, it was stationed at Strasbourg. In A.D. 69, in the first battle of Cremona, it had a bad time with a newly organized legion, I Adiutrix, which even captured its eagle. It did better later in the battle but then made another mistake by not siding with Vespasian. Unlike four other legions that made this mistake, it was not disbanded but sent instead to Bonn. In A.D. 85 or 86, it replaced its old enemy, I Adiutrix, at Mainz. Again, in A.D. 89, it seems to have revolted, and this time it was packed off to Pannonia, probably Hungary. It participated in the Emperor Domitian's campaign against the Sarmatians and seems to disappear entirely in A.D. 92, either being destroyed in battle or disbanded.

Augustus

XXII DEIOTARIANA

The name of this legion is unique. Deiotarus, king of Galatia in what is now northern Turkey, organized his troops along Roman lines, and during the time of Caesar, he fought campaigns in support of Caesar. The kingdom was bequeathed to Augustus in 25 B.C., and this formation was accepted into the Roman army as a legion bearing the name of its Galatian founder.[5] Prior to 8 B.C., it was stationed three miles from Alexandria, at Nicopolis, a fortress built for it by Augustus. Reports show it continuing in Egypt in A.D. 23, 70, and 110. In A.D. 119, it was still in Alexandria (Nicopolis) sharing a camp with II Cyrenaica which was sent off the following year to Arabia. In A.D. 132, the Jewish war started, and XXII Deiotariana disappears, either destroyed in combat or abolished. It seems to have spent virtually its entire career in Egypt in or near Alexandria.

Notes

1. See Tacitus, *The Annals,* I, 40–44.
2. See Dio, LXII, Loeb, VIII, pp. 83–105.
3. See Lewis and Reinhold, pp. 507–509.
4. For an account of the Varian disaster, see Dio, LVI, Loeb, VII, pp. 39–51.
5. For more on the founding of this legion, see Parker, pp. 271–272.

Nero

CHAPTER VI

First-Century Formations

During the reigns of Tiberius, who succeeded Augustus, and of Gaius, his successor, no new legions were added to the twenty-five that existed in the year A.D. 14. The first new legions, XV and XXII Primigenia, were raised by Claudius, A.D. 41–54, presumably to make available other legions for the invasion of Britain. Nero raised two more and Galba another, named for himself. Vespasian added three more but also disbanded or removed four more from the list. Finally, the Emperor Domitian added one more. V Alaudae was destroyed, however, in A.D. 92, and XXI Rapax disappears about the same time. Thus, at the close of the first century, there were in being twenty-eight legions. It is the nine organized by these five emperors that will be discussed in this chapter.

Fortuna

XV PRIMIGENIA

This legion has a very short and undistinguished record. Organized by Claudius,[1] A.D. 41–54, it and XXII Primigenia were the first legions to come into being since the death of Augustus in A.D. 14 and were named for the goddess of fortune, Fortuna Primigenia. Claudius, with his plans for the invasion of Britain, needed them to replace on the Rhine legions being withdrawn for the British campaign. In A.D. 43, it was stationed on the upper Rhine near Mainz. From there, it was moved to Bonn. In A.D. 69, it joined the revolt against Otho and then found itself being besieged at Vetera (Xanten) by Civilis, who was attempting to form his Gallic empire. It surrendered to Civilis, earning the keen displeasure of Vespasian, who shortly thereafter crushed the Civilis revolt. It appears that in A.D. 70 Vespasian disbanded its remnants and also three other legions, I Germanicus, IV Macedonica, and XVI Gallica. All four legions were considered to have disgraced themselves.

Plutarch, in his life of Galba, writes, ''Most say that he (Galba) was killed by a soldier of Legion XV.'' As the other Legion XV, Apollinaris, was at this time in the East, it may be assumed that XV Primigenia could claim the murder for one of its men.

Domitian

XXII PRIMIGENIA

Formed by Claudius at the same time as XV Primigenia, this legion had a long and distinguished record. The goddess Fortuna Primigenia, for whom it was named, smiled on it. Domitian honored it with the additional title *pia fidelis Domitiana,* and it had at one time as its commander a future emperor. When organized, it was first dispatched to Mainz where it shared a camp with IV Macedonica. This was about the year A.D. 39 when Claudius was preparing for his conquest of Britain.

It went off to Italy in A.D. 69 to fight at the first battle of Cremona and help Vitellius defeat Otho's legions and establish himself as emperor. He didn't last long, and a few months later the legion found itself on the wrong side of the second battle of Cremona when the forces of Vespasian defeated those of Vitellius. The legion was then packed off to Pannonia, probably Hungary. Vespasian must have thought well of the legion, for he did not abolish it as he did four others that had opposed him. In A.D. 70, it was back in the Rhine Valley, this time probably at Nijmegen in what is now the Netherlands. The base troops that had remained behind on the Rhine when the legion was in Italy seem to have been wiped out by Civilis during his abortive uprising.

Some time after A.D. 84 the double camp at Xanten was replaced by a stone fortress and occupied by this legion.

In A.D. 89, a civil war broke out when Antonius Saturninus, governor of upper Germany, led a revolt against the Emperor Domitian. The legions of the upper Rhine followed him, but those of the lower Rhine, including XXII Primigenia, retained their loyalty to the emperor. Maximus, the governor of lower Germany, led his troops against Saturninus and won a major victory against serious odds. The engagement took place on the plain of Andernach between Bonn and Coblenz, and German forces that had planned to cross the Rhine to help Saturninus were foiled by a sudden thaw of the frozen river. Later, when Domitian arrived on the scene, he rewarded the loyal legions with the additional title of *pia fidelis Domi-*

tiana. In view of the fact that Saturninus, whose head was sent to Rome, had financed the revolt by using the banks of two legions, Domitian issued an edict setting a limit on how much money could be held in such banks. At this time, he also forbade stationing of two legions in a single camp.

In A.D. 110, reports show the legion in upper Germany and in 120 stationed at Mainz. While stationed in Germany, it had as its commander Didius Julianus, who was to be emperor in A.D. 193.

A commemorative slab found on the Antonine wall in Scotland hints that about A.D. 193 an element of the legion was used in helping in the construction of the wall. There has come to light no other evidence of the legion having detachments in Britain. The last report we have on the legion is for the year A.D. 241, two centuries after its founding, when it was still stationed at Mainz.

I ITALICA

Nero

The Emperor Nero liked to do things in the grand manner and developed toward the end of his time as emperor a project to conquer the Caucasus, apparently with some thought of occupying the steppes of southern Russia. At the same time, he conceived another plan, to conquer Ethiopia. For the latter campaign, he started to concentrate troops in Alexandria, and as one of the steps in preparing for the Caucasus venture, he organized a new legion, I Italica. This legion was to be special, and he set up a requirement that all its members be from Italy and all over six feet tall. He called it "The Phalanx of Alexander the Great." Before either venture could be launched, Galba and Otho and the praetorian guard all joined in a revolt, and in January of A.D. 69 Nero committed suicide.

Organization of I Italica, which started in September A.D. 67, had been completed, and the legion was concentrated at Lyon in southern France. It got swept up in the Vitellian campaign against Otho, and when Vitellus was, in turn, overthrown by the forces of Vespasian, I Italica was sent off to Novae, near present-day Svishtov, on the Danube in Bulgaria. This general area south of the Danube was the Roman province of Moesia.

The legion, according to reports, seems to have remained based at Novae or nearby Oescus for the next century, for we have reports of it there in A.D. 79, 93, 96, 106, 110, and 138. The last report at hand is for the year A.D. 200 when it is still reported in lower Moesia, presumably Novae.

It must not be assumed that the legion had a sedentary life, for throughout this period the tribes north of the Danube were almost continuously at war with the Romans. We have reports of it participating in the Dacian war of A.D. 87, the wars with the Sarmatians, and, during Trajan's time, A.D. 98–117, at least two more wars with the Dacians.

Trajan

VII GEMINA (GALBIANA)

Before leaving Spain to serve his seven months as emperor in Rome, Sulpicius Galba, having been governor of Spain for eight years, raised a new legion, which bore his name, VII Galbiana. Its birth date is given as June 10, A.D. 68. It went to Rome with the new emperor, and after a short stay there was posted on the Danube at Carnuntum east of Vienna near Kaimbura. It was but a short time before the new emperor, Otho, called it back to Italy for his unsuccessful defense against Vitellius at the first battle of Cremona. Then it was back to the Danube again under the orders of Vitellius. It had now been depleted, reorganized, and renamed VII Gemina.

Vespasian, who was still in Alexandria, chose Antonius Primus, the commanding officer of VII Gemina, to organize forces for the takeover of the empire from Vitellius. Undoubtedly still smarting from the defeat at the first battle of Cremona, the legion strongly supported Vespasian and, with other like-minded legions, won the second battle of Cremona. This resulted in the death of Vitellius, December 20, A.D. 69, and the elevation of Vespasian. After this victory, it was "back again" to the station on the Danube.

Once things settled down, Vespasian apparently made a decision to transfer VII Gemina back to Spain, from whence it had come. On its way back from the Danube in A.D. 73 or 74, it was held for several years as part of the army of the upper Rhine but continued thereafter to Spain and was stationed at Leon, a communications center of major importance.

In A.D. 89, a revolt broke out in Germany with Antonius Saturninus, governor of upper Germany, having himself proclaimed emperor by the two legions stationed at Mainz. His revolt was short-lived,[2] but at its start, VII Gemina was summoned from Spain by the Emperor Domitian. Although it moved very rapidly, under the command of the future Emperor Trajan, it arrived too late for the decisive battle on the plain of Andernath. It then returned to Spain with the thanks of the emperor.

Records of the legion after this move are limited. It is reported in Spain in A.D. 110, 120, 138, 161, and 196. The last report available is for the reign of Caracalla when it is still stationed at Leon.

For whatever it is worth, the Notitia mentions an element of the legion in Spain in the fourth century.

Pertinax

I ADIUTRIX

Nero, in preparing for his projected campaign in the Caucasus, ordered the formation of two new legions, one to be organized using sailors of the Roman fleet. Not only was this source of soldiers unusual, but that they were not Roman citizens was unprecedented. Membership in the legion provided to its soldiers a means of getting this citizenship. The name given to the legion was I Adiutrix, meaning that it was a reserve or supporting legion "in addition to others." It had not been fully organized when Nero died, and some of its members met the Emperor Galba on his entrance to Rome and petitioned heatedly for certain privileges that were denied. Some sort of rioting took place, and several of the members of the legion were killed.[3] As a result, when Otho overthrew Galba, some months later, I Adiutrix was much on the side of Otho. There is a report that Galba had actually ordered the legion decimated for failing to respect his authority.

The legion was strong in support of Otho and fought with distinction for him at the first battle of Cremona. Although Otho's forces finally lost the battle, and Otho subsequently his life, I Adiutrix showed great ability, particularly for a new and untried outfit. In a head-to-head battle with experienced XXI Rapax, it gave the veterans a very bad time and actually captured their eagle. Before the battle was over, XXI Rapax counterattacked with equal success, but I Adiutrix had established itself as a legion to be reckoned with.

In A.D. 70 and again in A.D. 83, the legion was reported on the Rhine. It, with XIV Gemina, built and occupied a new stone fortress at Mainz. In A.D. 85 or in A.D. 86, it was replaced at Mainz by its old enemy XXI Rapax when it was called by Domitian to Pannonia (Hungary).

Briefly under Trajan, about A.D. 106, it provided occupation troops in Dacia (Romania). In A.D. 108, it is reported at Brigetio, near present-day Komaron, in Hungary on the Danube, about forty-five miles northwest of Budapest. About this time, it was honored with the additional title of *pia fidelis,* indicating special loyalty to Trajan during some unidentified revolt. In A.D. 115, elements of the legion were taken by Trajan for his Armenian campaign, but the legion is still reported at Brigetio in A.D. 120 and in A.D. 138. In A.D. 173, under the command of the future Emperor Pertinax, it is credited with liberating from invaders Raetia and Noricum in present-day Austria. In A.D. 216, it participated in the Mesopotamian campaign of Caracalla. In A.D. 228, it was back on its post on the Danube, presumably at Brigetio.

The legion was last in the limelight about the year A.D. 259 when the Emperor Gallienus brought it from the Danube to Italy to help repulse an invasion by the Alemanni near Milan. Although part of a relatively small army, estimated at ten thousand men, it administered a crushing defeat to the much larger host of invaders. This is the last identification at hand for the legion.

Caracalla

II ADIUTRIX

Like I Adiutrix, this legion was organized from sailors of the fleet, from Ravenna this time, and its name implies support or reserve status. Organization was not completed till March A.D. 70 when Vespasian incorporated it into the nine-legion army he dispatched to put down the revolt being led by Civilis in Germany. After a short stay in Germany, on the lower Rhine, it was transferred to Britain, first stationed at Lincoln and then at Chester where it was reported in A.D. 79. It was taken from Britain probably in A.D. 86 and sent to the Danube front where it was more needed. It is reported to have taken part in Trajan's Dacian campaign of A.D. 86–88 and to have fought the Suebi in A.D. 97. In A.D. 95, the nineteen-year-old future Emperor Hadrian was a tribune of the legion. In A.D. 105, the legion is identified at Singidunum (Belgrade), and in A.D. 106 it seems to have moved up the Danube to Aquincum near Budapest. It played a part in Trajan's Parthian campaign (Persia) in A.D. 113 and then returned to Budapest. In A.D. 120 and 138, it continued to be reported here in Pannonia. In A.D. 162 and 163, it participated in another Parthian war, this time during the opening years of the reign of Marcus Aurelius. In the following year, it was returned to Budapest, but in A.D. 173 there is a report of its being at Trencin, Czechoslovakia, deep in the lands of the Marcomanni, sixty miles north of the Danube. After its successful part in the Marcomanni war, it was once more returned to Budapest in A.D. 193. The last report is for the year A.D. 216 when the Emperor Caracalla took it to Mesopotamia for his Parthian campaign. It appears to have been returned then to Budapest.

IV FLAVIA FIRMA

Vespasian

Vespasian, the first of the three emperors of the Flavian family, gave the family name to two of the legions he raised in the year 70. Among the four legions he had disbanded were XVI Gallica and IV Macedonica, and he used their numbers and probably some of their better soldiers in forming XVI Flavia firma and IV Flavia firma.

The latter was sent first to Burnum, modern-day Knin in Dalmatia, Yugoslavia. In A.D. 86, the Emperor Domitian brought it north and east for the war in Dacia, present-day Romania. Dalmatia, which had revolted in the first decade, was now quiet, and no legion was required there. The situation was different in Dacia, which had been causing trouble since the last century B.C. Raiding had been taking place over the Danube from the north all along the river from Vienna to the Black Sea. In A.D. 85–86, under their powerful leader Decebalus, they were again raiding over the Danube into Moesia and were driven back. The Romans pursued into Dacia but there suffered a serious defeat. In A.D. 88, however, Domitian's legions, presumably including IV Flavia firma, won a major victory in Dacia itself. The legion was then stationed at Belgrade (Singidunum) and undoubtedly took part in Trajan's successful war against the Dacians in A.D. 106 that resulted in Dacia becoming a Roman province. Reports show it on the Danube in A.D. 108 and in A.D. 120. It appears to have remained in this part of Moesia either at Belgrade or fifty miles away at Viminacium during the whole next century. In A.D. 337 there is a report of the continued existence of the legion with a detachment in Upper Moesia.

Macrinus

XVI FLAVIA FIRMA

Like IV Flavia firma, XVI Flavia firma was organized by the Emperor Vespasian in the year A.D. 70 and bore his family name, with the quality of being steadfast added. Unlike IV Flavia firma, which was first stationed in Europe, XVI Flavia firma was initially established at Satala in northeast Turkey twenty miles north of present-day Erzinkin. Its history seems to be bound up with the East, whereas its sister legion was always in Europe. It formed a part of Emperor Trajan's army that made, in A.D. 114–115, the grand sweep down the Tigris and Euphrates rivers.[4]

Trajan then established it at Samosata, now known as Samsat, in southeast Turkey. This is a key crossing of the Euphrates and had to be secure for any Roman operations in the area. During Hadrian's time, A.D. 117–138, it and VI Ferrata were together the garrison of Samosata. In A.D. 138, it was reported in Syria, but shortly thereafter it was back in Cappadocia, presumably at Samosata again.

In A.D. 197 and 198, it was part of the Emperor Severus's army that captured Ctesiphon, the Parthian capital on the Tigris River south of present-day Baghdad. In A.D. 216, it was again taken from Samosata, this time by Caracalla for his Parthian campaign in Mesopotamia.

It was undoubtedly part of the army that the Emperor Macrinus put together in A.D. 217 for his not-very-successful Parthian campaign. In A.D. 218 the emperor was killed when resisting, from his base in Antioch, the forces of Heliogabalus, who then became emperor.

The Emperor Alexander Severus had it in his ten-legion concentration when he based himself in A.D. 231 at Antioch for his retaliatory campaign against the rising power of Persia. Under its new and effective leader, Ardashir, Persia was challenging the empire. This campaign was not a success for either side, and the records do not show in which of the columns the legion marched.[5]

Again the legion returned to Samosata, but in A.D. 260 it was probably withdrawn to western Turkey. Specific historical references to XVI Flavia firma now cease.

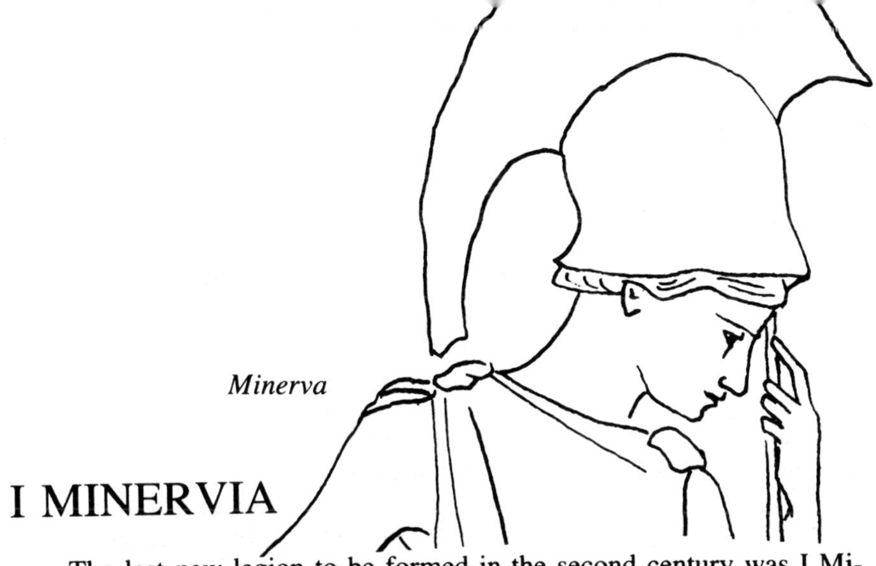
Minerva

I MINERVIA

The last new legion to be formed in the second century was I Minervia, organized by Domitian in A.D. 83 and named for the emperor's patron goddess. The requirement for more troops on the Danube for the Chatti Campaign had caused the withdrawal of XXI Rapax from Bonn, and the new I Minervia took its place there.

In A.D. 89 the legion showed its loyalty to the emperor when the Governor of the upper Rhine, Saturninus, led a revolt using the two legions at Mainz, XIV Gemina and XXI Rapax, which had returned from the Danube. I Minervia with the other lower Rhine legions supported the emperor, and the revolt was short-lived. As a result, I Minervia received the additional honorific title *pia fidelis Domitiana*.

Other results of this uprising was the prompt transfer of XXI Rapax back to the Danube, to be followed there two years later by its unsuccessful comrade in arms, XIV Gemina. It was at this time that Domitian issued the edict against stationing in the future two legions together in the same camp. He also set at this time the limit on the amount of money that could be held in the legion banks, for Saturninus had used the banks of the two Mainz legions to finance his uprising.

In the winter of A.D. 102–103, I Minervia was moved to the Danube for Trajan's campaign against the Dacians in what is now Rumania. The future emperor, Hadrian, commanded the legion in Trajan's successful handling of the Dacians, which resulted in the suicide of their commander and the capture of their capital.

It was, indeed, a fortunate legion that had a commander such as Hadrian. If the reader will permit an anachronism, Hadrian was a Renaissance man. Greek scholar, athlete, soldier at fourteen, administrator, legion tribune at nineteen, legion commander at thirty, and, unfortunately, a pederast. He was emperor at forty-two and a poet of some ability. His quips are still quoted by the scholars.[6] He, unlike other emperors, died a natural death in A.D. 138 at the age of sixty, just having composed a poem about his soul.[7]

By A.D. 110, the legion had been returned to the Rhine and was again reported at Bonn in A.D. 120 and again in lower Germany in A.D. 138, presumably at Bonn.

In A.D. 162, it made its only expedition to the east when the Emperor Marcus Aurelius used it in his highly successful war against the Parthians. The Parthian leader, Osroes, had taken over control of Armenia and had scattered the somewhat less energetic eastern legions while capturing much of Syria. Marcus Aurelius took his Rhine and Danube legions to Syria and completely defeated Osroes, restoring the Roman position in that part of the world to its high point under Trajan.

By A.D. 166, the legion was back at its base at Bonn and was reported still in upper Germany in A.D. 200 and 231. In the last-mentioned year, it was given a rough time keeping in hand the Franks, who were pushing hard on the Rhine frontier. This is the last firm date for I Minervia.

Notes

1. Authorities differ as to whether XV Primigenia was organized by Claudius or by Gaius.

2. See XXII Primigenia, above, p. 61.

3. See Plutarch, p. 1278.

4. See III Gallica, above, p. 33.

5. See XII Fulminata, above, p. 37.

6. In *Lives of the Later Caesars,* 15–11, the poet Florus is quoted as saying in a poem that he doesn't want to be Caesar and be among the Britains and in the frosty weather of Scythia. Hadrian replies in a poem that he does not want to be Florus in the taverns and cookshops and having to put up with the round insects.

7. In *Lives of the Later Caesars,* 26–1, the poem of the dying Hadrian is quoted, asking where the spirit, the guest and companion of his body, will go and commenting that it will go without its usual jest.

CHAPTER VII

Second-Century Formations

During the second century, seven new legions were organized, two at the start of the century by Trajan, two in midcentury by Marcus Aurelius, and three just at the close of the century by Septimius Severus. By the beginning of the third century, however, two Augustan legions, IX Hispana and XXII Deiotariana, disappear from the reports. Thus, in the first decade of the third century, there were thirty-three legions in being, four more than at the start of the second century.

During the last part of the second century and the beginning of the third, there continued the shift, mentioned in chapter V, of legions from the west and north to the east and south. An examination of the order of battle maps for A.D. 130 and A.D. 230, on pages 103 and 105, shows that although the number of legions had increased by two, the number on the Rhine had decreased from five to four. The number on the Danube had, during the same period, increased from ten to twelve, and in the Middle East, from eight to ten.

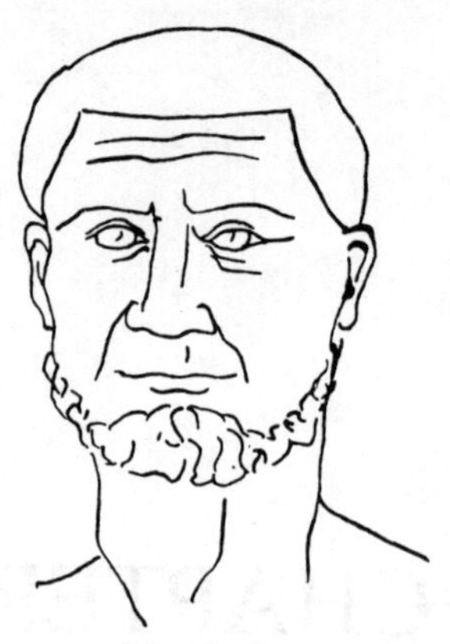
Maximinus

II TRAIANA

When the Emperor Trajan had settled things in Rome, he looked to the problem of Dacia that he inherited from Domitian and Nerva. His first move was to raise two new legions, one of them II Traiana, to bear his name. Dacia (Romania) had a strong leader, Decebalus, who united the forces of his kingdom. It required two campaigns before Dacia could be conquered. II Traiana was certainly part of the second campaign, for we have a firm date of A.D. 105 for the legion being in the area. In A.D. 110, it is reported in upper Moesia (Bulgaria), presumably holding one of the strong points on the Danube on the south border of Dacia.

In A.D. 114, Trajan had the legion available for his Armenian campaign and his subsequent drive down through Mesopotamia to the Persian Gulf.[1] Records do not show what part the legion played in this campaign.

Further reports show the legion in Egypt in A.D. 120, 135, and 194 and at Nicopolis near Alexandria in A.D. 231. In A.D. 232, the commanding officer of the legion was Gaius Julius Verus Maximinus, a giant soldier who had risen from the ranks and achieved the rank of centurion under Caracalla. After commanding the legion, he went on to be supreme commander of the Roman armies. In A.D. 235, he became emperor and ruled for three years. During the seige of Aquileia, which had espoused the cause of the next emperor, Gordian, he was murdered in his tent by troops of II Parthica, which had been presumed supporting him.

In A.D. 232, when Maximinus was in command, the legion had a detachment in Syria that participated in a small and unsuccessful revolt. The last report of the legion is for the year A.D. 288 when Diocletian sent a major portion of it on the expedition from Tangier to Carthage to resecure that part of the empire.

Trajan

XXX ULPIA VICTRIX

The second of the two legions formed by Trajan in response to the needs of the Dacian war was given the number XXX, as it became, in A.D. 101, the thirtieth active legion in the army. Having given his dynastic name to the first legion he formed, II Traiana, he gave the family name, Ulpius, to the other, XXX Ulpia Victrix. The family had a distinguished military background, as Hadrian's father, M. Ulpius Trajanus, had commanded V Macedonica at the seige and capture of Jerusalem, and Trajan himself had commanded VII Gemina in A.D. 89 when he rushed that legion from Spain to the Rhineland to help Emperor Domitian put down Saturninus' revolt.[2]

After the Dacian campaign, it was reported at Sarmzegetus in Dacia (southwest Romania near Vulcan in Transylvania). In A.D. 110, it had moved south to upper Moesia, probably on the Danube. From there, it was taken by Trajan for his Armenian campaign, but by A.D. 119 the legion had been transferred to lower Germany at Vetera, present-day Xanten. There it replaced VI Victrix, which had been rushed to Britain where there had broken out a serious revolt.

At this point, records of the legion become almost nonexistent, but there is one report that shows it in upper Germany in A.D. 138 and a final report for A.D. 269 when it was still at Vetera, having retained its headquarters for a century and a half at the same location.

Marcus Aurelius

II ITALICA PIA

Marcus Aurelius, emperor of Rome A.D. 161–180, was a scholar, statesman, philosopher, athlete, author, and soldier. Unlike other emperors, he did not find it necessary to name new legions after himself.[3] Instead, he named them for his native land, and II Italica Pia (patriotic) was the first to be formed. It was called into being because of the desperate situation both on the Danube and in Parthia (Persia). Usually, Rome was able to arrange to have one war at a time, but Marcus inherited two at once. It is regrettable that he preferred to write philosophy rather than history, for there is very little recorded as to the legions during the last half of the second century.

II Italica was raised either in A.D. 165 or 166 at a time when the plague was sweeping through the Roman army. In A.D. 169, records show the legion in Germany on the Danube at Casta Regina (Regensburg). After A.D. 170, it is reported at Lauriacum near Linz, Austria, 100 miles to the east but still on the Danube. It probably served in the Parthian war, but dates do not seem to be available.

When Marcus Aurelius died at Vindobonna (Vienna) in A.D. 180, the legion was back on the Danube, at or near Regensburg. In A.D. 253, it is reported in Noricum, so it may well have been at its old base at Lauriacum. From Noricum, it was taken back to Italy by the Emperor Gallienus for his famous victory in A.D. 258 when, with outnumbered forces, he crushed the Alemanni near Milan. One can speculate as to the future actions of the legion, but no specific reports are at hand.

III ITALICA CONCORS

Marcus Aurelius

In addition to II Italica, Marcus Aurelius raised a second legion, III Italica Concors, the "concors" implying concordant, harmonious, and efficient.

When the organization of the legion was complete in A.D. 166, it was sent initially to the north of Italy to defend that frontier. Although records are very scarce, there is an indication that it was moved briefly to Germany and then went to the Parthian war with II Italica. By A.D. 170, it is reported in Noricum, which is roughly present-day Austria. There are two reports of it then being stationed at or near Regensburg where it seems to have remained till at least the opening of the third century.

In A.D. 216 the Emperor Caracalla took it to Mesopotamia for his Parthian campaign and presumably returned it to the Regensburg area thereafter.

Septimus Severus

I PARTHICA

Septimius Severus became emperor in A.D. 193 and immediately formed three new legions, I, II, and III Parthica,[4] the first and third of these being organized in the east in what was then Mesopotamia. I Parthica was first stationed at Nisibis near the modern town of Al Qamishli on Syria's northeast border.

Having been born at Lepcis Magna on the coast of present-day Libya, the emperor's native tongue was Punic, and he was the first emperor to whom Latin was a foreign language. After a very brief stay in Rome, he departed for Syria where some legions had indicated support for C. Pescennius Niger's bid to be Rome's ruler. Earlier Severus had himself commanded a legion in Syria.

The army that Severus put together may well have included the new I Parthica. In any event, the defeat and death of Niger were accomplished in A.D. 194. Byzantium held out, however, against Severus until A.D. 196 when he captured and destroyed the city and demolished its then-famous wall. The principal inhabitants were put to the sword. In A.D. 198, the legion is again reported at Nisibis.

In A.D. 231, a report shows the legion having moved to Singara, a hundred miles to the southeast and about thirty miles from present-day Al Mawsil. Although it can be presumed to have participated in the successive wars with Parthia, no actual record is at hand.

The last firm report concerning the legion is for the year A.D. 231 when it was still at Singara and participated with III Parthica in a mutiny that cost the governor of the province his life.

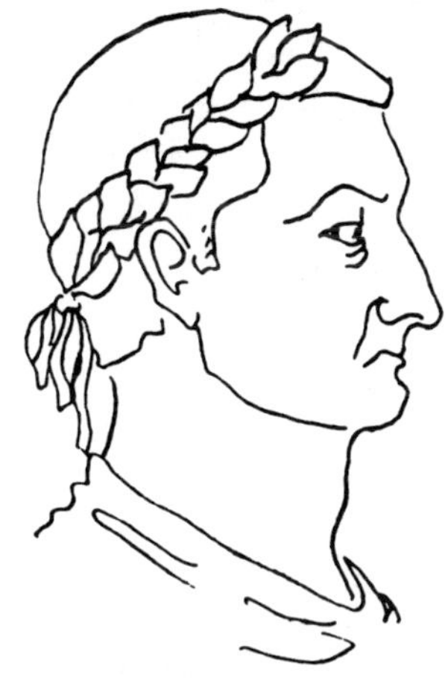

Gordian

II PARTHICA

Organized by the Emperor Septimus Severus in A.D. 193, this legion, unlike its sisters I and III Parthica that were formed in the east, was established at Albano some twenty miles south of Rome. It would appear that the emperor wished to have troops that he could count on near the capital.

It is not known whether it was part of the force used by Severus when he defeated Clodius Albinus, the candidate for the purple who brought his legions from Britain and Gaul. The battle took place in A.D. 197 at Lyon, and although there is no evidence to prove it, the strategic location of II Parthica would certainly indicate that it should have been there.

The legion is listed as one of those that participated in Caracalla's Parthian campaign in Mesopotamia in A.D. 216. In A.D. 218, it joined forces with III Gallica at Raphanaea in supporting Elagabalus's bid for the empire. Its success in the battle south of Antioch won for it the title *pia fidelis felix aeterna*.

In A.D. 238, it is back in Italy and participated in the seige of Aquileia, near Monfalcone on the Gulf of Venice. The then emperor, Maximinus, was attempting to capture the city that had gone over to the Gordian claimant to the empire. Troops from II Parthica murdered Maximinus and his son in their tent, thus winning praise from the Gordian followers.

The legion appears in a somewhat more favorable light in A.D. 258. It had returned to its base at Albano but was taken from there by the Emperor Gallienus, who put an army together to hold against the Alemanni. This German force had penetrated well into Italy. At Milan, he completely defeated the Alemanni, although he had fewer troops. The enemy was virtually destroyed, and the remnants were chased out of Italy. After this, the legion presumably returned to Albano, but there are no more firm reports.

Valerian

III PARTHICA

The story of III Parthica is much the same as that of its sister legion, I Parthica. Both were formed by the Emperor Septimius Maximus in A.D. 193, and both had their first station at Nisibis in Mesopotamia closer to the Tigris than the Euphrates. Both were participants in the successful wars of Severus against the Persians.

Shortly before A.D. 231, the legion moved with I Parthica to Singara, 100 miles to the south, and there it helped to prove the correctness of the injunction of Emperor Domitian, issued about A.D. 84, against stationing two legions in the same place at the same time. The two Parthica legions revolted in A.D. 231 and killed the governor of the province. Unlike the case of I Parthica, we do have a general report concerning this legion's demise. In A.D. 235 it was a functioning legion, but in A.D. 284 it no longer existed.[5] Lacking available records, it must be assumed that both it and I Parthica were destroyed when Mesopotamia was overrun by the Persians in the campaigns of A.D. 230, 241, 256, and A.D. 260. It was in this last year that the Emperor Valerian was captured at Edessa by the Persian leader Shapur. He was to die in captivity shortly thereafter. In A.D. 241, Nisibis, the birthplace of the legion, had fallen to the Persians.

By the time III Parthica disappears, the structure of the Roman legion was changing fast. The emperors were more and more using detachments from various legions rather than the legions themselves as fighting units. By the year A.D. 300, few legions remained intact with the morale and capabilities of the legions of the first two centuries.

Notes

1. See III Gallica, above, p. 32.

2. See VII Gemina, above, p. 64.

3. The family of Marcus Aurelius was originally Spanish, from Uccubi, a chartered town, says the Augustan History. His greatgrandfather had come to Rome and had the rank of praetor in the Senate.

4. See *Cambridge Ancient History*, vol. 12, p. 24. Webster, on p. 110, disagrees, showing II Parthica raised in the east, but subsequent events dispose of Webster's location. Lewis and Reinhold, on p. 491, have all three Parthica legions probably in being during the reign of Antonius Pius, which is highly unlikely.

5. See Edward N. Luttwak, *The Grand Strategy of the Roman Empire From the First Century A.D. to the Third* (Baltimore, Md.: Johns Hopkins University Press, 1976; paperback, 1979) p. 175.

EPILOGUE

After the third century, the legions no longer were preeminent as defenders of the empire. Diocletian, toward the end of the century, increased the number of legions from roughly thirty to over sixty, but at the same time the strength of the legions was greatly reduced, as were their capabilities for countering enemy incursions.

Although the names and numbers of the majority of the legions survived well into the fourth century, they were no longer employed in the same manner as in the great days of the empire. The change was gradual, but, in due course, the role of the legion as a counterattacking force was almost eliminated. During the first three centuries, the legions were used as a distributed reserve. The auxiliary troops manned the frontiers, and only when they could no longer handle a situation was a legion called. Likely as not, it not only restored the situation but invaded the enemy area and cut to pieces the former raiding force. During these earlier centuries, the legions were slid along the support area to the east or south as most needed. They constituted a close-up mobile reserve.

In the fourth century, central reserve forces were established, made up largely of cavalry and usually under the direct control of the emperor. Although many writers considered this an improved concept, it may well have been exactly the reverse. The still stirrup-less cavalry could move faster than the legions but had normally much greater distances to cover. They were not as easily moved by water and, in some cases, would have to take land routes when legions would have gone by a direct water route.

An examination of map 4 shows sixteen legions along the Rhine-Danube front. With auxiliary units manning the defenses, the legions were usually free to move rapidly to endangered areas. For short moves, legions could cover considerably more than twenty miles a day.

A critical situation at Bonn on the Rhine could call forward three legions in three days and a fourth in another four. Similarly, at Carnuntum on the Danube, three legions could make the march in two days, and a third legion could be there three days later. It is unlikely that a central reserve would even have heard about the problem by the time the legions would have reached the area. Even on the much less strongly held Euphrates front, an incursion between Samosata and Melitene could be met within three days by two legions, and a third would be on the scene in eight days. It would be weeks, or more likely months, before a central reserve could appear.

It is also questionable whether forces made up largely of cavalry were as effective as the legions. Enemy tactics had not changed, and one recalls that III Gallica had little trouble in A.D. 69 in destroying a force of nine thousand Sarmatian cavalry. Similarly, a ten-thousand-man army assembled by Gallienus from I Adiutrix, II Parthica, II Italica, VIII Augusta and the Praetorian Guard defeated and destroyed, in A.D. 258–259, a much larger force of invading Alemani at Milan.

The great strengths of the legions lay in their training, morale, and leadership. Training obviously suffered when the legions were broken up and used in defensive positions. Attack training must have suffered most severely. Morale was generally high in the first three centuries, and the legions believed strongly in themselves. Also, in the early years, the high rank of the legion commanders led to self-confidence in the troops, for they knew that they were a power in the empire. So did the emperor. Ten legion commanders used their legions as springboards and became, themselves, emperors. The soldiers and centurions, serving an emperor who had himself served in or commanded a legion, were conscious of their might. To what degree the reduction of the legions' power and effectiveness was a result of imperial fear of the legions is not clear. This fear, however, must have played a significant part in the weakening of the empire.

In the centuries when the legions were at their prime, there was a very special esprit not based on pay, skill, training, or technical leadership. A willingness to die for the empire existed among the troops and their leaders. This we find in other great military organizations: British, French, German, and American. There was, in addition, a personal honor that required success and did not permit failure. This is not unlike that found on occasion among Japanese troops.

Emperors, centurions, and legion commanders killed themselves when they failed in their tasks, The centurion Sempronius Densus, alone defending his emperor from the rebel troops in Rome, was not an isolated example. Neither was the suicide of Poenius Postumas, acting commander of II Augusta, when he failed to bring the legion into the battle against Boudicca. This concept of the honor of the soldier has been found in other countries at other times but seldom over such a length of time as it was found in the legions, three hundred years.

APPENDIX I

Emperors of the First Three Centuries

B.C.

27 Augustus

A.D.

14 Tiberius
37 Gaius (Caligula)
41 Claudius
54 Nero
68 Galba
69 Otho
Vitellius
Vespasian
79 Titus
81 Domitian
96 Nerva
98 Trajan
117 Hadrian
138 Antonius Pius
161 Marcus Aurelius
180 Commodus
193 Pertinax
Didius Julianus
Septimus Severus
211 Caracalla
217 Macrinus
218 Elagabalus
222 Alexander Severus
235 Maximinus
238 The two Gordiana
Pupienus and Balginus
Gordian III
244 Philip
249 Decius
251 Gallus
253 Valerian
Gallienus
259 Gallienus, alone
268 Claudius II
270 Quintillus
Aurelian
275 Tacitus
276 Probus
282 Carus
283 Carinus and Numerian
284 Diocletian (Maximian associated with him, 286)
305 Constantius and Galerius

APPENDIX II

Emperors Serving with Legions

Vespasian	commanded II Augusta in A.D. 43
Titus	had command of at least one legion during the Jerusalem campaign
Trajan	commanded VII Gemina in A.D. 89 in northern Spain
Hadrian	commanded I Minervia in Dacia about A.D. 106 and had previously been a military tribune with II Adiutrix
Pertinax	commanded I Adiutrix and had previous service as a military tribune
Didius Julianus	commanded XXII Primigenia in Germany
Septimus Severus	commanded IV Scythica in Massiam (probably Masyaf in western Syria)
Maximinus	prefect of II Traiana in A.D. 232 in Nicopolis, Egypt
Aurelian	commanded X Gemina
Probus	commanded X Gemina

APPENDIX III

Stations of the Legions

Localities listed below are those mentioned in the text where legions were based for one or more years. Temporary stations and sites of battles are not included. The Roman name is shown in parenthesis when appropriate. A Roman name with an asterisk indicates that the base was in the vicinity of the named city or town but not necessarily at precisely the same place. Letters and numbers following a Roman name indicate approximate direction and distance in miles from the present town or city.

BRITAIN	
Caerleon	II Augusta
Chester (Deva)	II Adiutrix
	XX Valeria Victrix
Gloucester (Glevum)	II Augusta
	XX Valeria Victrix
Lincoln (Lindum)	II Adiutrix
	IX Hispana
Newcastle	VI Victrix
York (Eburicum)	VI Victrix
	IX Hispana
FRANCE	
Lyon (Lugdunum)	I Italica
Strasbourg (Argentorate)	II Augusta
	VII Augusta
	XXI Rapax

SPAIN	
Leon	VII Gemina
NETHERLANDS	
Nijmegen (Noviomagnus*)	IX Hispana
	X Gemina
	XXII Primigenia
GERMANY	
Bonn (Bonno)	I Minervia
	XV Primigenia
	XXI Rapax
Cologne (Colonia Agrippina)	I Germanica
	V Alaudae
	XX Valeria Victrix
	XXI Rapax
Mainz (Mogantiacum)	I Adiutrix
	IV Macedonica
	XIV Gemina
	XV Primigenia
	XVI Gallica
	XXI Rapax
	XXII Primigenia
Neuss (Novaesium)	VI Victrix
	XVI Gallica
	XX Valeria Victrix
Regensburg (Castra Regina)	II Italic
	III Italica
Trier	I Germanica
Xanten (Vetera)	VI Victrix
	V Alaudae
	XXI Rapax
	XXII Primigenia
	XXX Ulpia
SWITZERLAND	
Windisch (Vindonissa)	XI Claudia
	XIII Gemina
	XXI Rapax

ITALY	
Albano	II Parthica
Aquileia	XI Claudia
Trieste (Emona* NE 50)	XV Apollinaris
AUSTRIA	
Linz (Lauriacum*)	II Italica
Vienna (Vindobonna)	X Gemina
	XIII Gemina
	XV Apollinaris
Hainburg (Carnuntum*)	VII Gemina
	XII Fulminata
	XIV Gemina
	XV Apollinaris
HUNGARY	
Budapest (Aquincum*)	II Adiutrix
Komeron (Brigetio*)	I Adiutrix
CZECHOSLOVAKIA	
Trencin*	II Adiutrix
YUGOSLAVIA	
Belgrade (Singidunum)	II Adiutrix
	IV Flavia firma
Knim (Burnum)	IV Flavia firma
	XI Claudia
Pozarevac (Viminacium*)	IV Flavia
	VII Claudia
Marburg (Poetovio)	XIII Gemina
Zagreb (Siscia*)	IX Hispana
RUMANIA	
Alba Iulia (Apulum*)	XIII Gemina
Baila (Troesmis*)	V Macedonica
Balta (Mehadia*)	XIII Gemina
Cluj-Napoca (Potaissa*)	V Macedonica
Vulcan, Transylvania (Sarmizegethusa NW*)	XXX Ulpia

Location	Legion
BULGARIA	
Archar (Ratiaria*)	XIII Gemina
Knizha (Oescus*)	V Macedonica
Silistria (Durostorum)	XI Claudia
Svishtov (Novae)	I Italica
TURKEY	
Antioch	IV Scythica
(Cyrrhus)	IV Scythica
	X Fretensis
Birecik (Zeugma*)	IV Scythica
Erzinkin (Satala* N20)	XV Apollinaris
	XVI Flavia firma
Malatya (Melitene)	V Macedonica
	XII Fulminata
Samsat (Samosata)	VI Ferrata
	XVI Flavia Firma
SYRIA	
Al Mausil (Singara* SW30)	I Parthica
	III Parthica
Al Qamishli (Nisibis*)	I Parthica
	III Parthica
Damascus (Danabe* NE)	III Gallica
Latakia (Laodicia)	VI Ferrata
Tartus (Raphanaea* E30)	III Gallica
ISRAEL	
Jerusalem	X Fretensis
JORDAN	
Irbid (Bostra*)	VI Ferrata
	III Cyrenaica
EGYPT	
Alexandria	II Traiana
	III Cyrenaica
	V Macedonica
	XV Apollinaris
	XXII Deiotariana
Luxor (Coptos* N)	III Cyrenaica

ALGERIA

Batna (Lambaesis*)	III Augusta
Tebessa (Ammaedara* NE20)	III Augusta

APPENDIX IV

Lifelines of the Legions

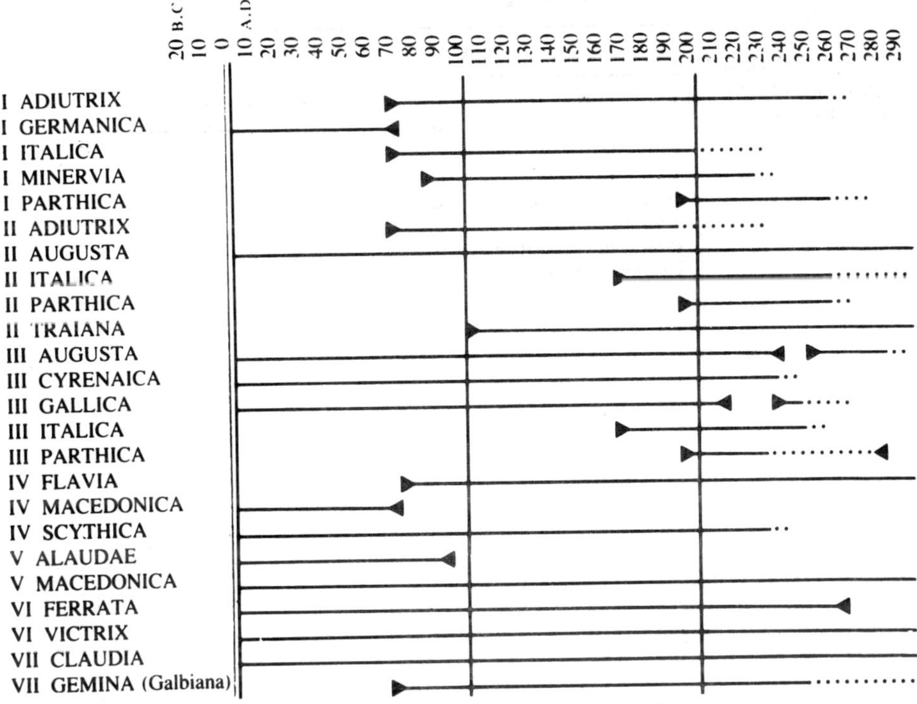

Lifelines of the Legions (continued)

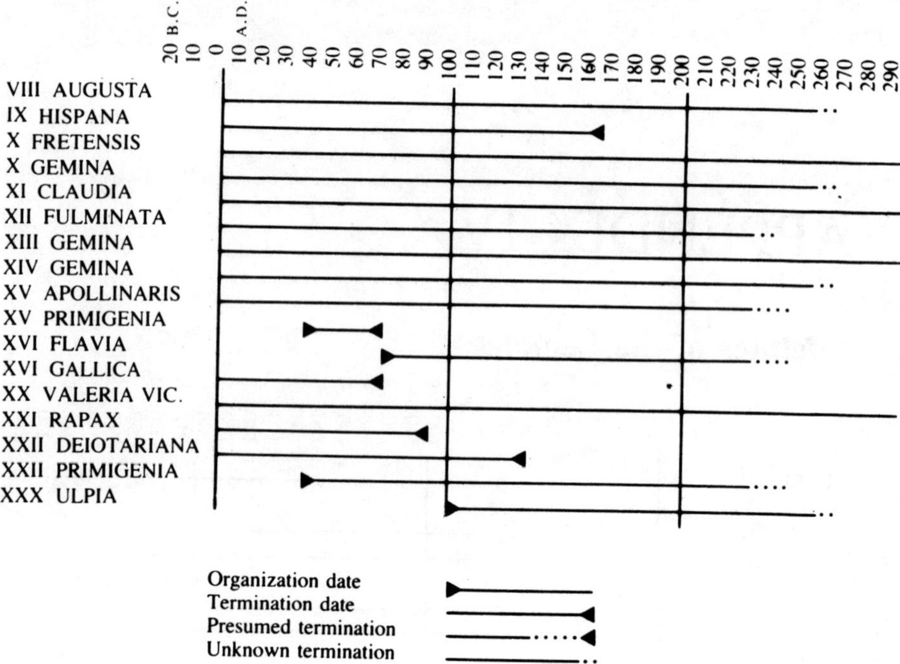

Organization date
Termination date
Presumed termination
Unknown termination

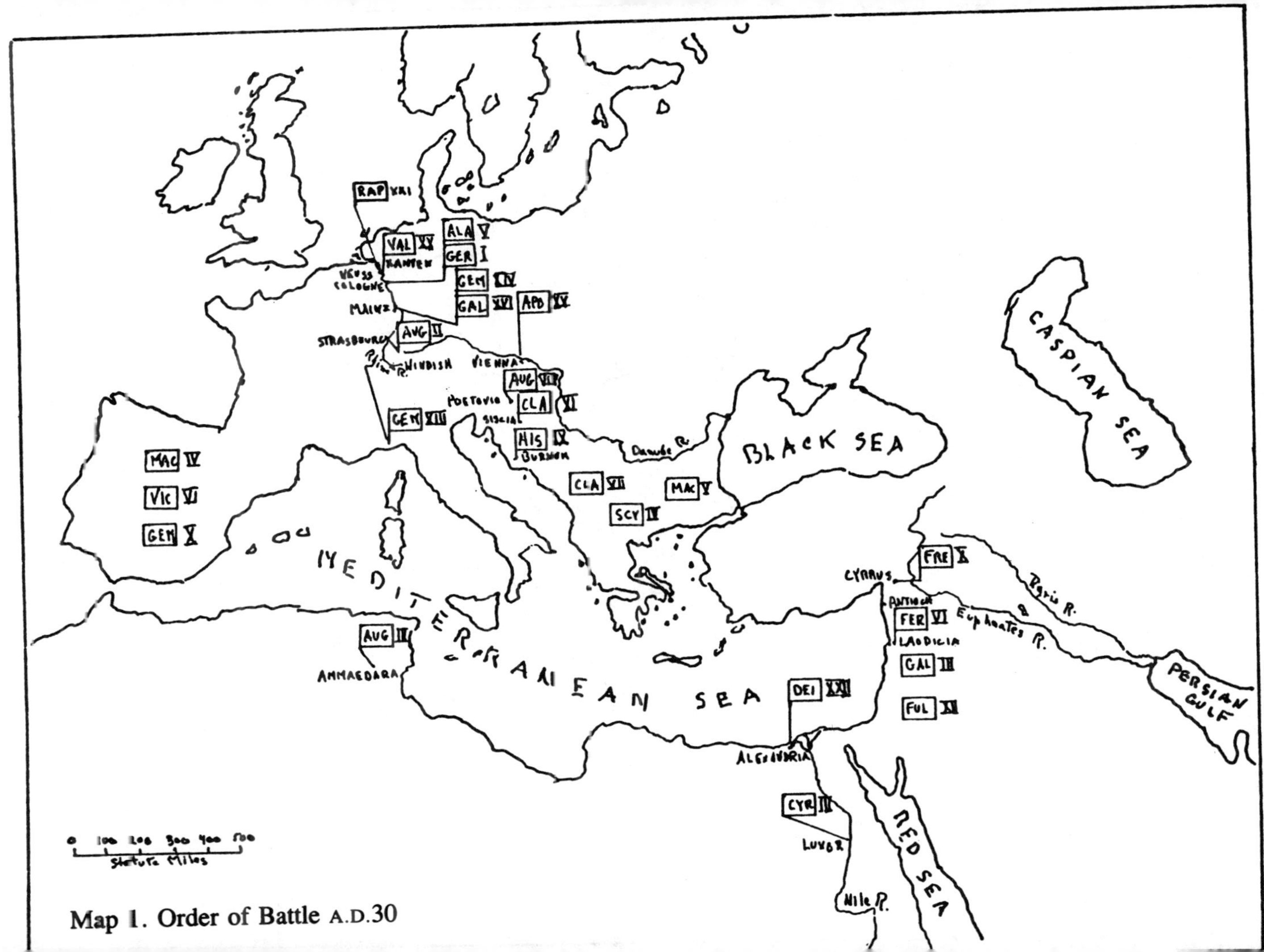

Map 1. Order of Battle A.D.30

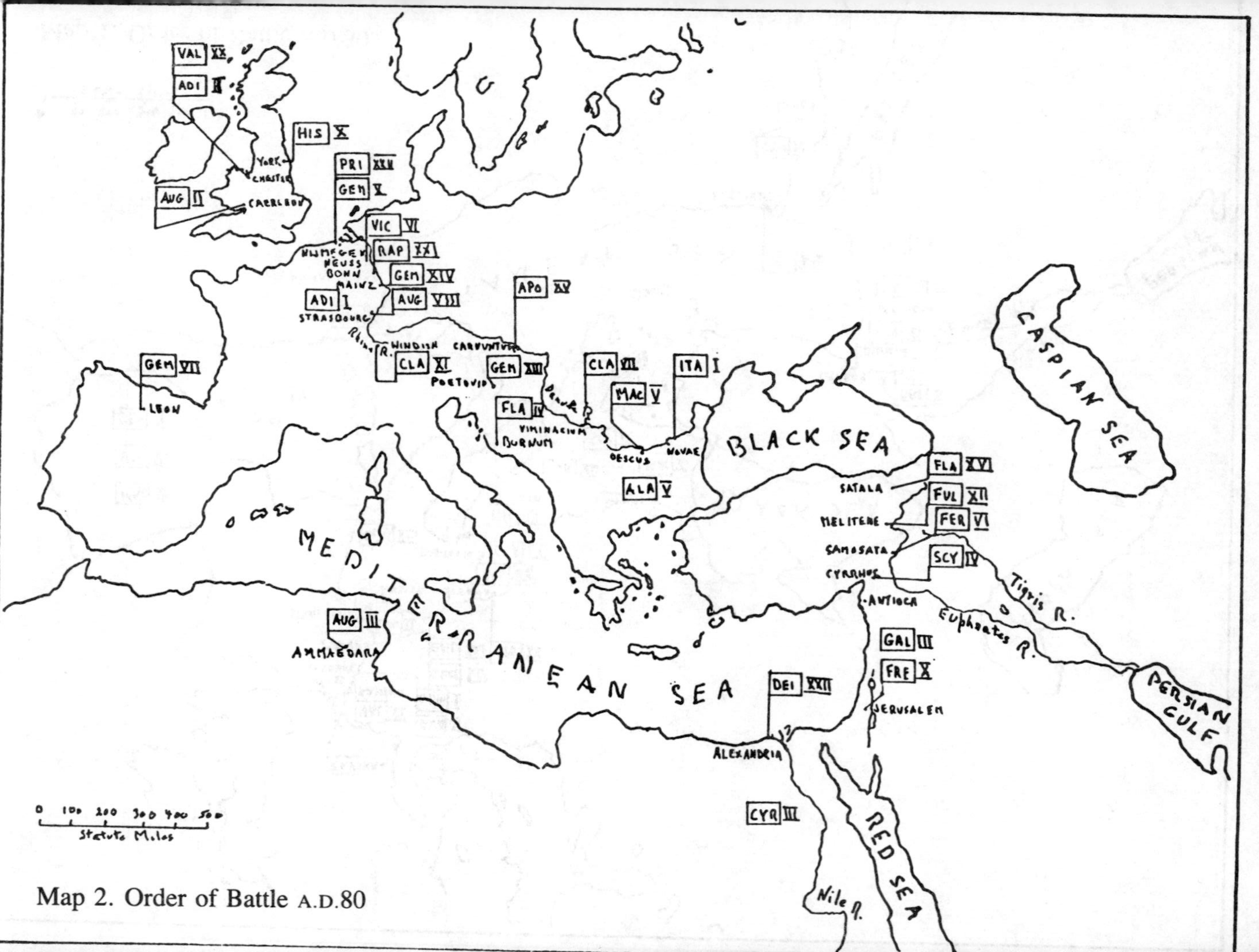

Map 2. Order of Battle A.D.80

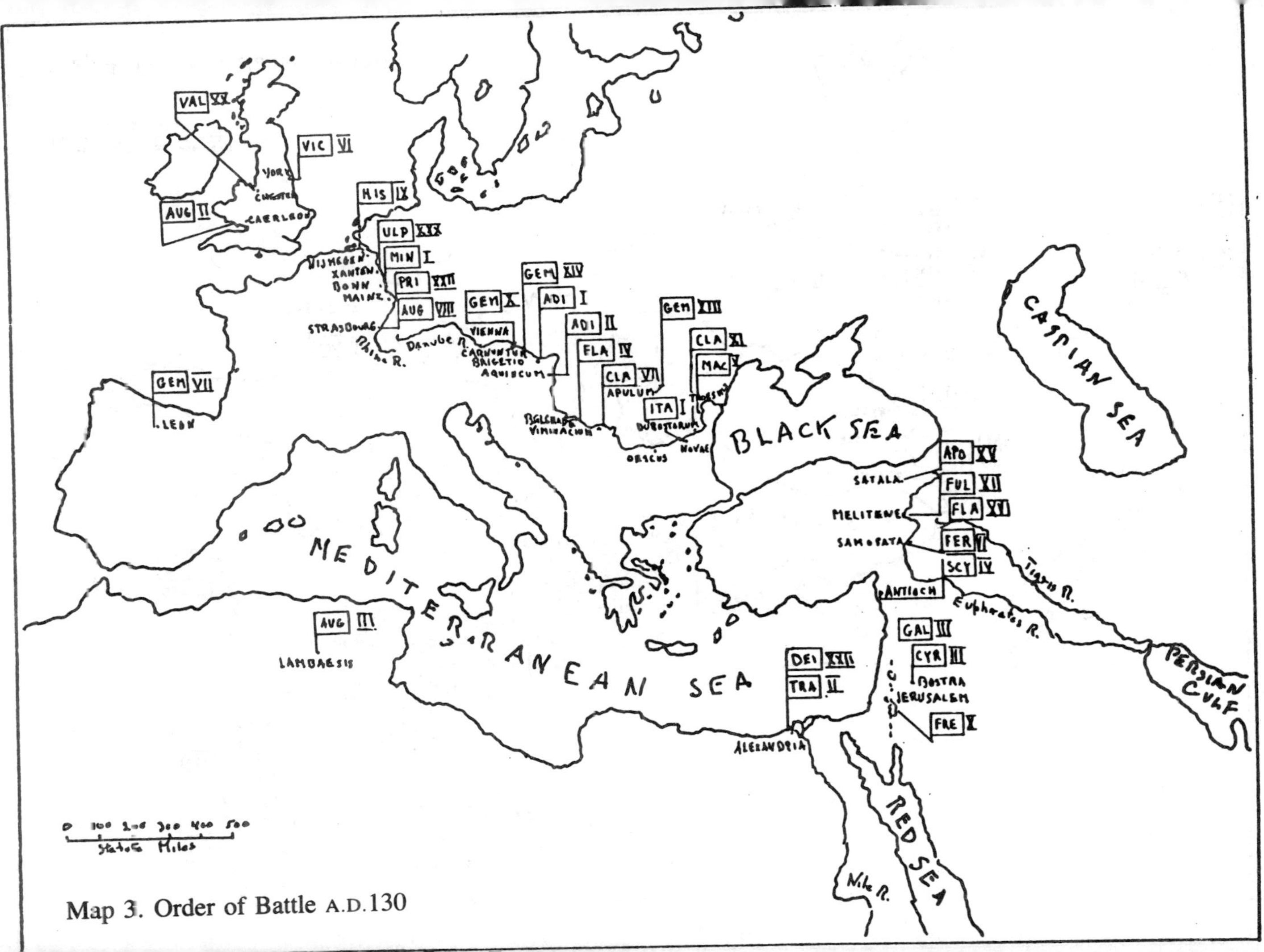

Map 3. Order of Battle A.D. 130

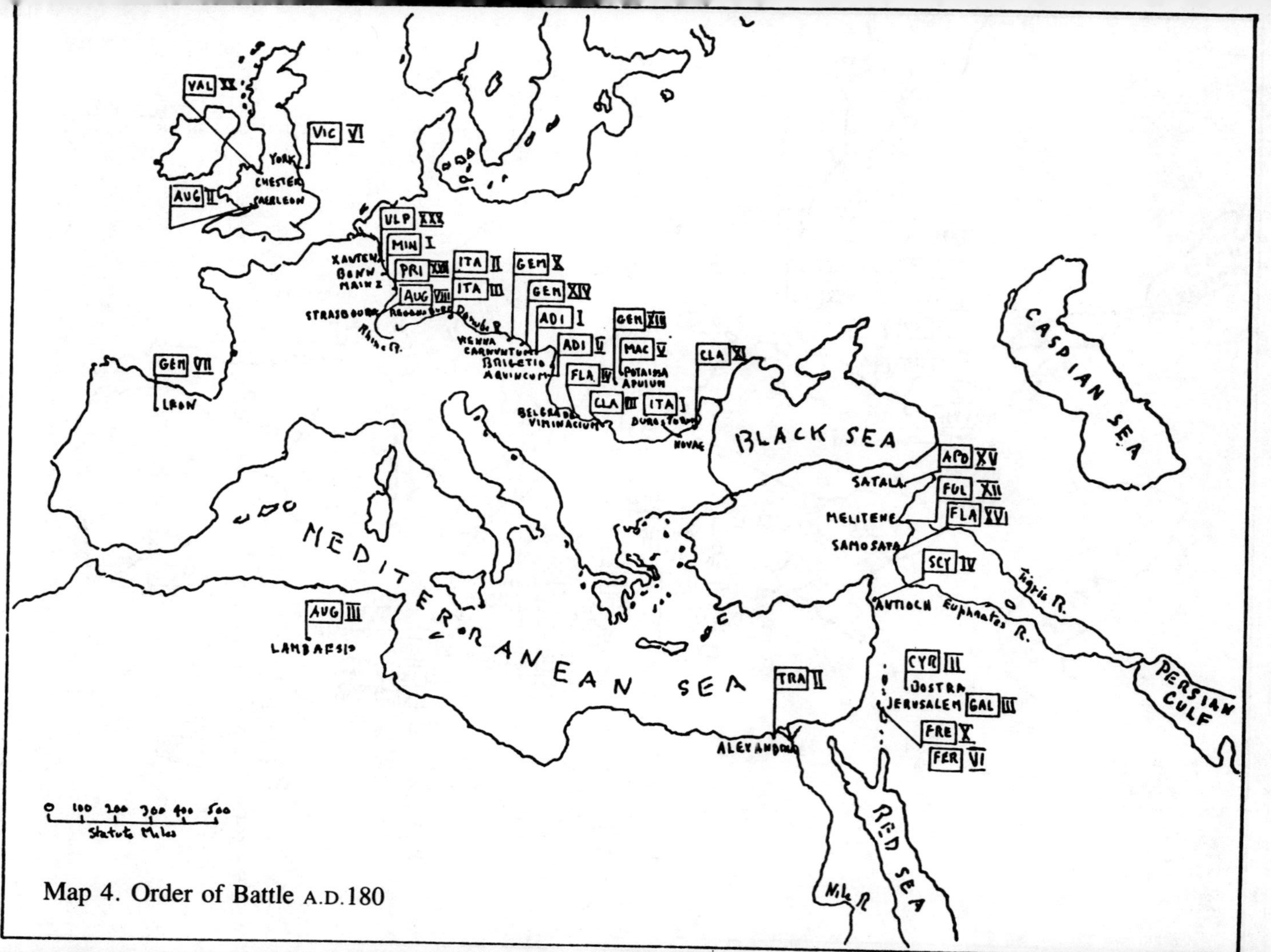

Map 4. Order of Battle A.D. 180

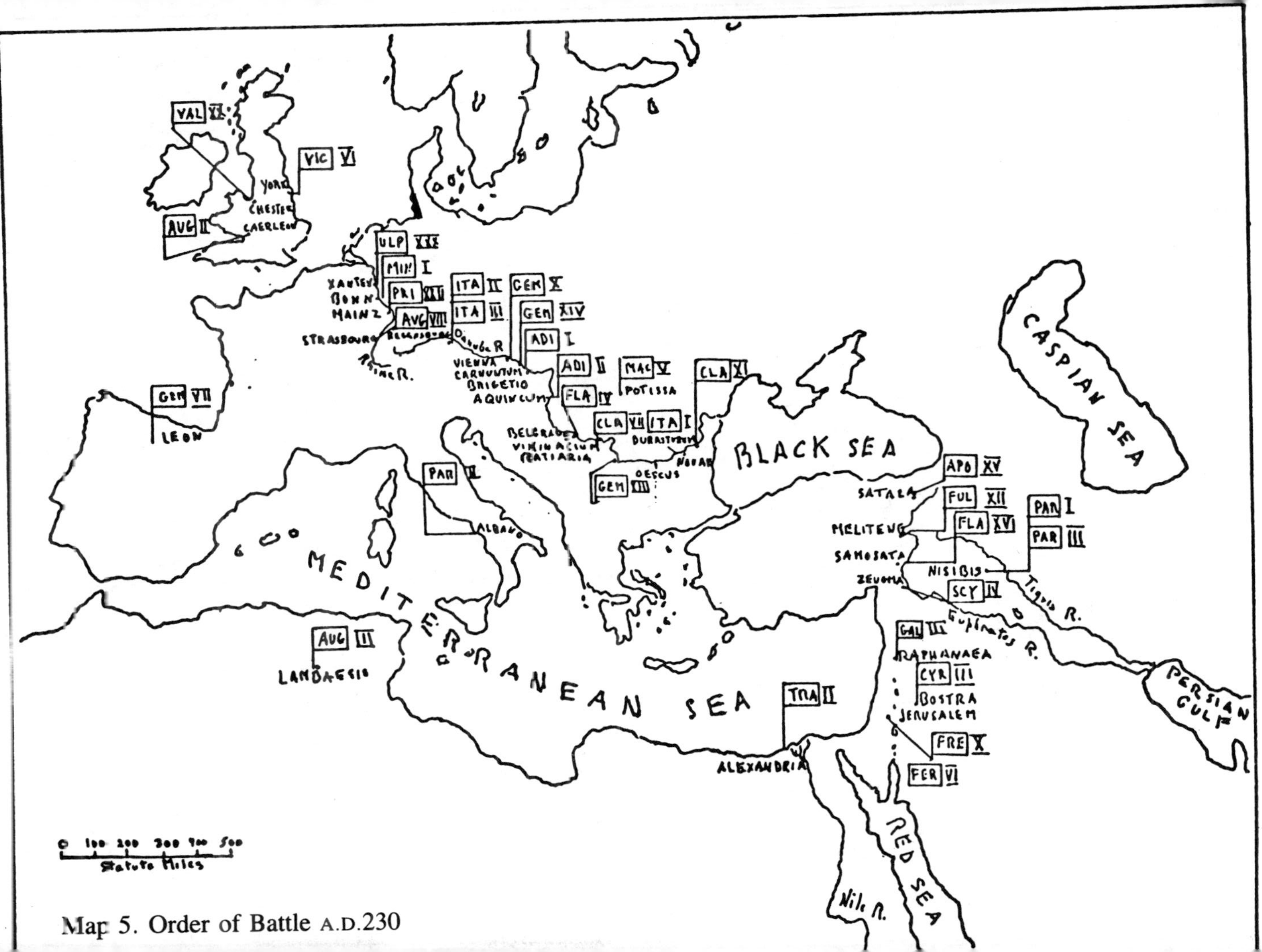

Map 5. Order of Battle A.D.230

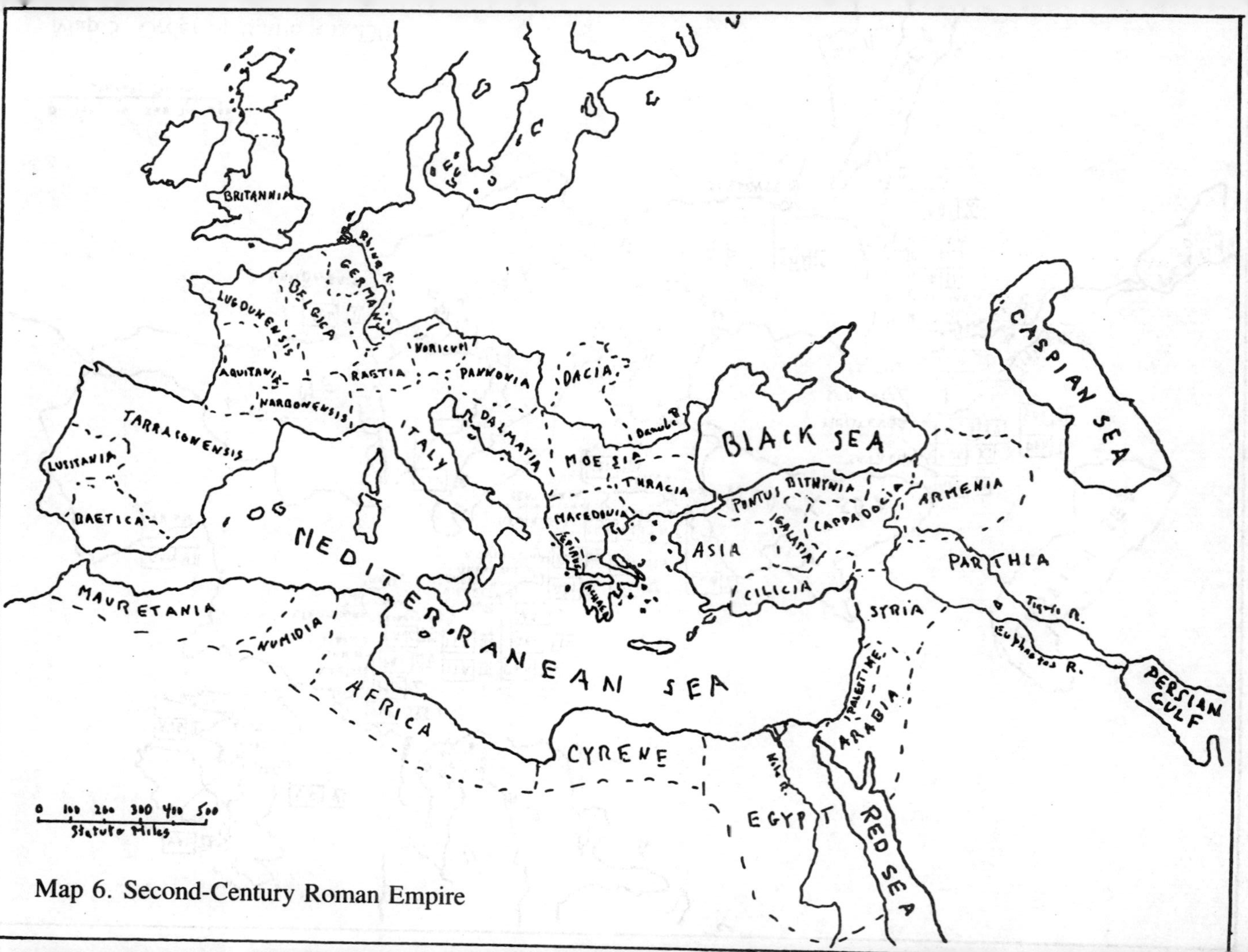

Map 6. Second-Century Roman Empire

BIBLIOGRAPHY

The books listed below, which were used in the preparation of this study, are all available through bookstores and, with the exception of Webster, are all in print in 1979. Except for the *Cambridge Ancient History,* which is to be had at most libraries, they are reasonably priced, and fourteen of them, marked with an asterisk, are sold in paperback editions.

A. Classic sources, all available in translation, are:

**Lives of the Later Caesars*. The first part of the *Augustan History*. Translated and with an introduction by Anthony Birley. New York: Penguin, 1976. Although this *Lives* has an unenviable reputation for inaccuracy, it provides some information concerning a dozen legions. It is skillfully edited with entertaining, if caustic, footnotes. An excellent index lists legions.

Dio, Cassius. *Roman History*. Translated by Earnest Cary. Cambridge, Mass.: Harvard University Press, 1924, nine volumes. Volumes seven and eight include Dio's books LVI-LXX and cover the first 138 years of our era. Dio in these books only mentions legions by name twice, but the detailed descriptions of battles and the conduct of troops are valuable. There are amusing and slightly salacious descriptions of the private lives and conduct of the emperors and their ladies. The footnotes are useful and entertaining.

Herodian. *The History*. Translated by C. R. Whittaker. Loeb Classical Library, two volumes. Cambridge, Mass.: Harvard University Press, 1969. Herodian rarely mentions a legion by name, but the superb and extensive footnotes provide valuable late-second-and early-third-century identifications. There is no index to legions.

*Josephus. *The Jewish War*. Translated with an introduction by G. A. Williamson. New York: Penguin, 1959. Josephus, who commanded the Jewish troops during much of the war, is a most unlikable character, but his book is precise history and contains invaluable descriptions of the organization and conduct of the Roman legions. After his capture, he became an enthusiastic supporter of the Romans against his own people. There are three maps.

*Lewis, Naphtali, and Reinhold, Meyer. *Roman Civilization Sourcebook II: The Empire* New York: Harper, Torchbook edition, 1966. This volume, with introduction and notes and a superb bibliography, is invaluable in its presentation in translation of a large number of historical documents pertaining to the Roman army and its legions.

Plutarch. *The Lives of the Noble Grecians and Romans*. Translated by John Dryden with revisions and introduction by Arthur Hugh Clough. New York: Modern Library, This great translation contains little about legions, but the lives of Galba and Otho throw much light on the period and the effect of the legions.

Scriptores Historiae Augustae. The entire *Augustan History,* both in Latin and translation, with an introduction by D. Magie; Loeb Classical Library, three volumes. Cambridge, Mass.: Harvard University Press, 1922. This supplements the above-mentioned *Lives of the Later Caesars* and contains the more inaccurate last half of the *History*. It is entertaining reading but filled with misinformation. Excellent notes and a superb index.

*Suetonius. *The Twelve Caesars*. Translated and with an introduction and notes by Robert Graves. New York: Penguin, 1957. This superb translation brings the first century A.D. to life with the stories of twelve emperors. Scandalous anecdotes abound, but there is still some valuable information and very enjoyable reading. Individual legions get no significant mention.

*Tacitus. *The Agricola and the Germania*. Translated and with an introduction by H. Mattingly and translation revision by S. A. Handford. New York: Penguin, 1970. Cn Julius Agricola was not only a great governor of Roman Britain but also the father-in-law of Tacitus, who gives him in this book a very favorable biography. There is an interesting and detailed account of the battle of Mons Graupius and of the campaign associated with it.

*Tacitus. *The Annals of Imperial Rome*. Translated and with an introduction and notes by Michael Grant. New York: Penguin, 1956. Although Tacitus was something of a prig, he provides in this volume helpful information about the doings of the legions during the first two-thirds of the first century. Legion facts are buried in stories of the emperors, but they are nonetheless valuable. There is an excellent index and good maps.

*Tacitus. *The Histories*. Translated and with an introduction and notes by Kenneth Wellesley. New York: Penguin, 1975. This contains much useful and precise information about the legions and dramatic detailed accounts of the first and second battles of Cremona. There are eleven maps, but, unfortunately, the index does not address itself to legions.

B. Modern histories, all in English, are:

*Birley, Anthony. *Life in Roman Britain*. London: Batsford Ltd., 1964; 1976. This has a good chapter on the Roman army in Britain, a useful map, and the index is helpful in dealing with seven legions.

*Blair, Peter Hunter. *Roman Britain and Early England, 55* B.C.–A.D. *871*. New York: Norton, 1963. The second, third, and fourth chapters contain interesting and useful information. Unfortunately, the author rarely mentions individual legions and does not mention them in the index.

*Breeze, David J., and Dobson, Brian. *Hadrian's Wall*. New York: Penguin, 1976; rev. ed., 1978. This is an exceptionally fine study, well researched and well presented. There is much information on the Roman army in Britain not found elsewhere. The index has numerous legion listings.

Cambridge Ancient History. Cambridge: Cambridge University Press, vol. X, 1934; vol. XI, 1936; vol. XII, 1939. These three volumes are without doubt the most useful studies of the Roman Empire and the Roman army for the period 44 B.C. to A.D. 325. There are numerous maps, and each volume has a superb index with more than one hundred individual legion listings.

Encyclopaedia Britannica. Cambridge: Cambridge University Press, 1911–1912. This world-famous eleventh edition is studded with classical data not found readily available elsewhere. There is a fine article in vol. XXIII, p. 471, on the Roman army and the legions by Francis J. Haverfield with an order-of-battle for the year A.D. 120.

*Goodrich, L. Carrington. *A Short History of the Chinese People*. New York: Harper Torchbooks, 1963. This volume is included only because of the mention, on p. 43, of Roman legionnaires confronting Chinese soldiers in Sogdiana.

*Grant, Michael. *The Climax of Rome*. London: Sphere Books, Ltd., 1974; Cardinal edition. Chapter 2 of this work is a discussion of the changes in the Roman army in the third century. It does not deal with specific legions but does provide a useful background. There are eight maps.

*Luttwak, Edward N. *The Grand Strategy of the Roman Empire From the First Century A.D. to the Third*. Baltimore, Md.: Johns Hopkins University Press, 1976; paperback, 1979. This provides an interesting discussion of the strategic problem and has some good order-of-battle information. There are fifteen maps; but the index does not list individual legions.

Mellersh, H. D. L. *The Roman Soldier*. New York: Taplinger Publishing Co., 1965. This provides general background.

Parker, H. M. D. *The Roman Legions*. New York: Barnes and Noble, first published by Oxford University Press in 1928 and reprinted in 1958 with corrections. This is a very thorough, if rather contentious, study of the legions up to the year A.D. 180. Appendix A is a unique and invaluable report on the origins and nomenclature of the Augustan legions. An excellent index provides easy reference to individual legions.

*Richmond, I. A. *Roman Britain*. Baltimore, Md.: Penguin Books, first published in 1955; 2nd ed., 1963. The first sixty-five pages present a good military history of the Roman occupation with six maps or charts. The index lists legions individually.

Scullard, H. H. *Roman Britain, Outpost of the Empire*. London: Thames and Hudson, 1979. This does not address itself specifically to legions, nor does the index make any reference to them. There are, however, some useful data on legionary fortresses and on the diet and life of the Roman soldier in Britain.

Webster, Graham. *The Roman Imperial Army of the First and Second Centuries A.D.* New York: Funk and Wagnalls, 1969. This is one of the most useful and thoroughly researched books on the Roman army. It covers the army in the field and in peacetime activities. There are good maps and charts and a superb index that follows individual legions by name from station to station.

INDEX
To Legion Histories

X	Gemina	51
XI	Claudia Pia Fidelis	52
XII	Fulminata	37
XIII	Gemina Pia Fidelis	53
XIV	Gemina Martia Victrix	54
XV	Apollinaris	25
XV	Primigenia	60
XVI	Flavia Firma	70
XVI	Gallica	55
XX	Valeria Victrix	14
XXI	Papax	56
XXII	Deiotariana	57
XXII	Primigenia	61
XXX	Ulpia Victrix	77